SOMETHING

ABOUT

EVERYTHING.

SAMUEL LEONARD.

"IT'S NOT POSSIBLE TO KNOW EVERYTHING ABOUT SOMETHING……….. BUT IT'S VERY POSSIBLE TO KNOW SOMETHING ABOUT EVERY THING"

Blaise Pascal.

DEDICATION.

Dedicated To my Hero Dad Mr. Benson Nchekwube Obodo.

THE NIGERIAN CIVIL WAR……….The Rise and fall of Biafra!

The Nigerian Civil War, also known as the Biafra War, took place from July 6, 1967, to January 15, 1970, and remains one of the most tragic episodes in Nigerian history. This conflict was primarily rooted in ethnic, political, and economic tensions that escalated into a full-blown armed conflict, resulting in significant loss of life and widespread suffering.

The roots of the Nigerian Civil War can be traced back to the amalgamation of Nigeria by the British in 1914, which brought together diverse ethnic and cultural groups under a single colonial administration. The subsequent struggle for political power, economic resources, and ethnic dominance set the stage for the eruption of violence.

The tensions between the Eastern Region, predominantly inhabited by the Igbo ethnic group, and the rest of Nigeria reached a breaking point when Lt. Col. Odumegwu Ojukwu, the military governor of the Eastern Region, declared the secession of Biafra on May 30, 1967.

The Eastern Region, dominated by the Igbo people, had a distinct history and identity. The Igbo, one of Nigeria's major ethnic groups, were known for their enterprising spirit and contributions to trade, commerce, and education. However, political power in Nigeria was often concentrated in the hands of other ethnic groups, particularly the

Hausa-Fulani in the north and the Yoruba in the southwest.

The post-independence period in Nigeria saw a struggle for political power and representation among the various regions. The Eastern Region, under the leadership of Dr. Michael Okpara, felt marginalized and underrepresented in the federal government. The control of resources, particularly oil revenue, became a contentious issue, further exacerbating regional disparities.

Lt. Col. Odumegwu Ojukwu, a charismatic and ambitious military officer, became the military governor of the Eastern Region in 1966. Ojukwu inherited a region facing political instability and ethnic tensions, with the Igbo people increasingly feeling marginalized. The political climate was further destabilized by coups and counter-coups that rocked Nigeria in the mid-1960s.

A significant turning point was the anti-Igbo pogrom that erupted in the northern part of Nigeria in July 1966. Thousands of Igbos were killed, leading to a massive influx of refugees into the Eastern Region. The failure of the federal government to protect the Igbo population deepened the sense of insecurity and alienation.

To address the escalating tensions, a meeting was held in Aburi, Ghana, in January 1967, between Nigerian leaders, including Ojukwu, to discuss the future of the country. The Aburi Accord aimed at finding a peaceful resolution to the political crisis but ultimately failed due to disagreements over its implementation.

Faced with increasing hostility and violence against the Igbo people, Lt. Col. Ojukwu declared the secession of the Eastern Region on May 30, 1967, establishing the Republic of Biafra. In his declaration speech, Ojukwu cited the need to protect the Igbo people from genocide and ensure their right to self-determination.

The outbreak of hostilities occurred when the Nigerian government, led by General Yakubu Gowon, rejected the secession of Biafra, leading to a series of military confrontations. This chapter will delve into the major battles, strategies employed, and the humanitarian crisis that ensued as the conflict escalated.

Battle of Nsukka (July 1967):

The first major confrontation occurred in July 1967 when federal forces attempted to recapture the town of Nsukka, a key location in the Eastern Region. The Biafra forces, under Colonel Chukwuemeka Odumegwu Ojukwu's command, successfully repelled the federal troops, establishing an early indication that the conflict would not be easily resolved.

Blockade and Economic Strangulation:

The Nigerian government implemented a blockade to cut off Biafra from external support, particularly in terms of weapons and supplies. This blockade, enforced by the Nigerian Navy, significantly impacted Biafra's economy and exacerbated the emerging humanitarian crisis.

Biafra Offensive into Midwest Region (August 1967):

In an attempt to expand the territory under Biafra control, Ojukwu ordered an offensive into the Midwest Region. However, the Nigerian government, supported by foreign mercenaries, successfully repelled the Biafra forces, reversing their territorial gains.

Battle of Asaba (October 1967):

The Battle of Asaba was a tragic episode where federal troops entered the town of Asaba, located on the western bank of the Niger River. Reports suggest that a significant number of civilians, primarily from the Igbo ethnic group, were massacred during the occupation. This event underscored the brutality and ethnic tensions that characterized the conflict.

Biafra Strategy:

Biafra adopted a defensive strategy, relying on guerilla warfare tactics and fortifications to resist federal advances. Colonel Ojukwu aimed to consolidate Biafra's territory and force the federal government into recognizing its secession.

Nigerian Strategy:

General Gowon's government pursued a strategy of "no-victor, no-vanquished," emphasizing the need for reconciliation and the reintegration of the

Eastern Region into Nigeria. The federal forces focused on offensive operations, seeking to regain control of Biafra territory and end the secession.

As the war continued, international actors became increasingly involved. The roles of these foreign nations, humanitarian organizations, and the impact of the conflict on civilians, including the infamous Biafra famine cannot be over emphasized.

Soviet Union and Eastern Bloc:

The Soviet Union, along with other Eastern Bloc nations, supported the Nigerian government with military aid, including weapons and training for federal forces. This support was driven by Cold War dynamics, with the Eastern Bloc aligning with the

Nigerian government as a counterbalance to Western influence in Africa.

Britain and the Commonwealth:

The United Kingdom, Nigeria's former colonial power, maintained a delicate position during the conflict. While publicly supporting Nigeria's unity, the British government faced criticism for its perceived role in supplying weapons to the Nigerian military. The Commonwealth also grappled with divisions over the appropriate response to the crisis.

France and Biafra:

France emerged as a key supporter of Biafra, providing diplomatic and military assistance. French officials, including President Charles de Gaulle, sympathized with Biafra's struggle for self-determination. French arms and relief aid were channeled to Biafra, contributing to the prolongation of the conflict.

International Committee of the Red Cross (ICRC):

The ICRC played a crucial role in providing humanitarian assistance to both sides of the conflict. They facilitated the evacuation of wounded soldiers, distributed medical supplies, and attempted to address the pressing humanitarian needs caused by the blockade. However, their efforts faced challenges due to the reluctance of the Nigerian government to allow unrestricted access to Biafra.

Médecins Sans Frontières (MSF):

MSF, also known as Doctors Without Borders, was founded in response to the Nigerian Civil War. Their medical teams worked tirelessly to provide healthcare in the midst of the conflict. MSF's efforts were particularly directed at addressing the severe malnutrition and health crisis affecting the civilian population, especially children.

Joint Church Aid (JCA):

JCA, a coalition of various church-based aid organizations, played a significant role in delivering relief to Biafra. They operated under challenging conditions, often facing obstruction from the Nigerian government. JCA's efforts focused on providing food, medicine, and support to the beleaguered Biafra population.

Biafra Famine:

Cause of Famine:

The Nigerian government's blockade, aimed at crippling Biafra economically, had severe consequences on the civilian population. The lack of access to food and medical supplies, combined with the disruption of agricultural activities, led to a severe famine. The Biafra famine is estimated to have resulted in the death of millions, with a disproportionate impact on children.

International Response to Famine:

Images of emaciated children and the scale of suffering in Biafra drew global attention. Humanitarian organizations intensified their efforts, and international public opinion pressured governments to act. The Biafra famine became a symbol of the catastrophic consequences of war and the urgency of providing humanitarian relief.

The international involvement in the Nigerian Civil War had profound implications for the conflict's dynamics, the provision of humanitarian aid, and the overall course of events. The Biafra famine, in particular, serves as a tragic testament to the complex interplay between geopolitical interests, humanitarian imperatives, and the human cost of armed conflict. As the war continued, international actors faced ethical dilemmas and challenges in navigating the complexities of the Nigerian Civil War.

The war officially ended on January 15, 1970, with the surrender of Biafra forces......A day to be remembered.......

Unconditional Surrender of Biafra:

On January 12, 1970, Colonel Chukwuemeka Odumegwu Ojukwu, the leader of Biafra, fled the country, and his deputy, Philip Effiong, announced the unconditional surrender of Biafra. This declaration effectively brought an end to the conflict, halting the violence that had ravaged the nation for over two years.

Reintegration Efforts:

In the aftermath of the war, the Nigerian government, led by General Yakubu Gowon, initiated efforts to reintegrate the Eastern Region, formerly Biafra, back into the country. The "3 Rs" policy—Reconciliation, Reconstruction, and Rehabilitation—was launched to rebuild the war-torn region and heal the wounds of the conflict.

No Victor, No Vanquished Policy:

General Gowon's government proclaimed a policy of "No Victor, No Vanquished" to emphasize national reconciliation. This policy aimed at fostering unity and encouraging a sense of shared responsibility for the rebuilding process. However, the effectiveness of this policy was debated, as deep-seated ethnic and regional tensions persisted.

The Indigenization Decree:

In an attempt to address economic disparities, the Nigerian government implemented the Indigenization Decree of 1972, which sought to transfer economic power from foreign-owned companies to Nigerian citizens. While this policy aimed at reducing economic imbalances, its implementation faced challenges and criticisms.

Ethnic and Regional Tensions:

Despite the government's efforts at reconciliation, ethnic and regional tensions lingered. The scars of the war, coupled with lingering perceptions of marginalization, contributed to a sense of distrust among Nigeria's diverse ethnic groups. This mistrust would continue to influence the nation's politics in the years to come.

Oil Revenue and Resource Allocation:

The control of oil resources remained a contentious issue post-war. The discovery of oil in Nigeria had contributed to economic imbalances, with the oil-producing regions, predominantly in the Niger Delta, seeking a fair share of revenue. Disputes over resource allocation would become a recurring theme in Nigeria's post-war political landscape.

Impact on Biafra's Identity:

The war profoundly impacted the identity of the Igbo people and the region that was formerly Biafra. The scars of the conflict, coupled with the economic and infrastructural challenges, shaped the socio-political trajectory of the Eastern Region. Efforts to rebuild and redevelop faced obstacles, impacting the region's development.

Military Rule and Governance:

The Nigerian Civil War set the stage for a series of military coups and countercoups, ultimately leading to a period of military rule in Nigeria. The post-war years witnessed political instability, with the military intervening in governance, exacerbating challenges in establishing a stable democratic system.

Impact on International Relations:

The Nigerian Civil War had implications for Nigeria's relations with other nations. While the conflict strained ties with countries that supported Biafra, efforts were made to mend diplomatic relationships in the post-war era. Nigeria's role in regional and international affairs evolved, shaping its standing on the global stage.

The end of the Nigerian Civil War marked the beginning of a complex and challenging period for the nation. The socio-political landscape of Nigeria was profoundly shaped by the war's aftermath, influencing governance, economic policies, and inter-ethnic relations for years to come. The scars of the conflict would continue to resonate, serving as a reminder of the importance of addressing underlying issues to build a united and stable nation.

BEHIND THE MINDS OF A SERIAL KILLER..........The untold story of Jeffrey Dahmer!

Serial killers have long fascinated and horrified society, and among the most notorious is Jeffrey Dahmer.

Jeffrey Dahmer's Background:

Early Childhood:

Jeffrey Lionel Dahmer was born on May 21, 1960, in Milwaukee, Wisconsin. His childhood appeared ordinary, but there were signs of disturbed behavior. Some experts believe that early incidents of cruelty to animals were early indicators of his later violent tendencies. He reportedly engaged in dismembering and mutilating animals, which some experts consider to be a red flag for potential violent behavior. The significance of such behavior lies in the hypothesis that cruelty to animals might be an early indicator or precursor to later violent acts against humans. This concept is often associated with the "triad" of behaviors, a theory suggesting that individuals who exhibit cruelty to animals, fire-setting, and bedwetting beyond a certain age may be more prone to violent behavior in the future.

The Making of a Killer:

"If peace can come through killing someone, then I don't want it", a statement made by Thomas Merton didn't go down well with Jeffrey as the whole world later realized. Dahmer's descent into criminality would mark the genesis of a monster we never knew....A lesson to be learnt!

Adolescence and Isolation:

Dahmer struggled with a sense of isolation and alienation from an early age. His family moved frequently during his childhood. Dahmer's parents went through a tumultuous divorce when he was a teenager. This period was marked by intense

conflict, and Dahmer's relationship with both parents was strained. His sense of abandonment and lack of emotional support from his family may have intensified his feelings of isolation. Dahmer's fascination with dissection and cruelty to animals during adolescence is often considered a warning sign. Many experts in criminal psychology suggest that such behavior can be an early indicator of deeper psychological issues and potential future violence.

Social Struggles:

Dahmer struggled socially during adolescence, reportedly feeling isolated and unable to connect with his peers. This sense of isolation likely contributed to his growing sense of detachment from others and difficulty forming meaningful connections.

Early Criminal Activities:

Dahmer's first brushes with the law involved incidents of indecent exposure and minor offenses. These early criminal activities foreshadowed the more heinous crimes he would commit later in life.

Military Service:

After high school, Dahmer enlisted in the U.S. Army. However, his military career was cut short due to alcohol-related issues and his increasing preoccupation with violent fantasies.

The First Murder:

Dahmer committed his first murder in 1978, shortly after leaving the army. Steven Hicks, a hitchhiker, became Dahmer's first victim. On June 18, 1978, Dahmer encountered Steven Hicks, who was hitchhiking to a rock concert. Dahmer offered him a ride and, during the journey, suggested they drink beer together. They eventually parked in a secluded area, where Dahmer bludgeoned Hicks to death with a dumbbell and then buried the remains. Hicks' disappearance initially went unnoticed, and the lack of a thorough investigation allowed Dahmer to avoid immediate scrutiny. This event however, marked the beginning of a horrifying series of murders that would continue for over a decade.

Luring Victims:

Dahmer targeted vulnerable individuals, often luring them to his apartment under the pretext of taking explicit photographs or offering money. His victims included young men from marginalized backgrounds.

Dahmer's preferred method of killing involved drugging his victims and then strangling them. His acts escalated in brutality, eventually involving dismemberment and necrophilia- a paraphilic disorder characterized by a persistent sexual attraction to corpses....

House of Horrors:

Dahmer's apartment in Milwaukee became a gruesome crime scene where he committed the murders and disposed of the victims' remains. The discovery of his "house of horrors" shocked the nation. Inside Dahmer's apartment, investigators discovered containers and bags containing human body parts, including heads, skulls, and other anatomical remains.

Refrigerator of Body Parts:

One of the most chilling aspects of Dahmer's crimes was his preservation of body parts in his refrigerator. The gruesome nature of his actions revealed the depths of his depravity. To dispose of the remains, Dahmer employed various methods. He experimented with acid to dissolve the bodies, using large drums in his apartment. This method aimed to destroy evidence and facilitate the disposal of human remains.

Polaroid Photographs:

Dahmer documented his crimes by taking Polaroid photographs of his victims before, during, and after their deaths. The chilling photographs provided a glimpse into the gruesome nature of his acts and served as incriminating evidence during the investigation.

Dahmer also attempted to create living zombies by injecting his victims' brains with chemicals. These experiments were part of his delusional and disturbed attempts to exert control over his victims even after their deaths.

Capture and Trial:

 A Failed Abduction:

Dahmer's capture was triggered by an attempted abduction that went awry. Tracy Edwards, a 32-year-old man, had encountered Dahmer at a bus station in Milwaukee. Dahmer eventually lured Edwards to his apartment with the promise of money for posing for nude photos. Edwards agreed, unaware of Dahmer's sinister intentions.

Once inside Dahmer's apartment, Edwards noticed disturbing evidences, including photographs of dismembered bodies and a foul odor. Realizing the danger he was in, Edwards resisted Dahmer's attempts to handcuff him.

During the struggle, Edwards managed to break free and fled from the apartment. He ran onto the street and flagged down two police officers, Robert Rauth and Rolf Mueller. Edwards led the officers back to Dahmer's apartment, providing critical information about the horrors inside.

Discovery of Evidence:

Upon entering Dahmer's apartment, the law enforcement agencies discovered a scene of horror, with evidence of dismembered bodies and human remains scattered throughout.... A clear indication that a monster had been on the loose. The arrest of Jeffrey Dahmer was immediate!

Trial and Conviction:

During the trial, Dahmer initially pleaded not guilty but later changed his plea to guilty by reason of insanity. The defense argued that Dahmer suffered from various mental disorders, contributing to his violent acts. This claim however, was not justified and Dahmer was set to face justice.

He was convicted of 16 murders and sentenced to 16 life terms in prison. The severity of his crimes shocked the public, and the trial highlighted the challenges of dealing with individuals exhibiting extreme psychopathy.....

End of Jeffrey Dahmer:

"There is no real ending. It's just the place where you stop the story..." Frank Herbert (American author).

On the morning of November 28, 1994, Dahmer and another inmate named Jesse Anderson were assigned to work-cleaning a bathroom. Christopher Scarver, another inmate on the work detail, attacked both Dahmer and Anderson with a metal bar that he had smuggled from the prison weight room. Scarver bludgeoned Dahmer and Anderson to death in the prison bathroom. Dahmer and Anderson both suffered fatal head injuries during the assault.

Both Dahmer and Anderson were pronounced dead at the hospital later that day. The motive behind

Scarver's actions, as well as the circumstances surrounding the attack, raised questions about the prison environment and the safety of inmates. Scarver later stated that he believed he was acting on orders from God to carry out the attacks. Scarver claimed that he was disturbed by the nature of Dahmer and Anderson's crimes and felt a sense of righteousness in carrying out the attack. This dark event drew the curtain close for the "Milwaukee Cannibal" thereby marking the end of a dark chapter in criminal history. It also spared the criminal justice system from further legal proceedings related to his case. However, the circumstances surrounding his death also sparked discussions about inmate safety and supervision within the prison system.

Lessons from Dahmer's Case:

Jeffrey Dahmer is dead to the world. He's come and gone like everyone else would. His legacy is all he has left of a true story that remains a trauma to the ears that heard them. His story serves as a reminder to parents all round the world, the importance of peace and unity in families-a key factor in shaping the hero or villain their children would become. The untold story of Jeffrey Dahmer sheds light on the darkest corners of the human psyche and forces our society to confront the chilling reality that such horrors can exist among us.

SNAKES IN THE CITY..........Before and After a Snake bite!

A snake bite refers to the injury caused by the bite of a snake. Snakes are found in various parts of the world, and while most species are non-venomous and pose little threat to humans, some can deliver venom through their bites, which can have varying degrees of toxicity. Snake bites are a significant public health concern in regions where venomous snakes are prevalent.

A Snake Bite!

The effects of a snake bite can vary significantly depending on several factors, including the species of snake, the amount of venom injected, the location of the bite, and the health and size of the person bitten. It's important to note that the majority of snake species are non-venomous and their bites are generally not life-threatening. However, bites from venomous snakes can have serious consequences. Here's a general overview of what happens when a snake bites:

Immediate Pain and Swelling:

In most cases, a snake bite results in immediate pain and swelling at the site of the bite. The severity of these symptoms can vary, but they are typically the first noticeable signs.

Local Tissue Damage:

Venomous snake bites may cause local tissue damage due to the toxic components in the venom.

This can result in bruising, blistering, and necrosis (death of cells or tissues) around the bite area.

Systemic Symptoms:

Depending on the type of snake and the potency of its venom, systemic symptoms may occur. These can include nausea, vomiting, weakness, dizziness, sweating, difficulty breathing, and changes in heart rate. In severe cases, a person may experience systemic shock.

Hematologic Effects:

Some snake venoms affect the blood-clotting mechanism. This can lead to spontaneous bleeding, both internally and externally. Hemotoxic venoms can cause issues such as hemorrhage, prolonged bleeding, and shock.

Neurotoxic Effects:

Certain snake venoms primarily target the nervous system. Neurotoxic effects can include paralysis, difficulty speaking, blurry vision, and difficulty swallowing. In extreme cases, respiratory failure may occur due to paralysis of the respiratory muscles.

Allergic Reactions:

Some individuals may experience allergic reactions to snake venom, which can exacerbate symptoms. Anaphylaxis, a severe allergic reaction, can be life-threatening and requires immediate medical attention.

Delayed Effects:

In some cases, the full effects of a snake bite may not manifest immediately. Certain snake venoms may have delayed or progressive effects, making it important for individuals bitten by potentially venomous snakes to seek medical attention even if initial symptoms are mild.

Long-Term Consequences:

Survivors of severe snake bites may experience long-term consequences, including tissue damage, scarring, and psychological trauma. Chronic pain, disability, and limb deformities can result from extensive tissue damage caused by venom.

After a Snake Bite!

Stay Calm:

It is important to remain as calm as possible. Panic can increase heart rate and promote the spread of venom in the bloodstream. Try to limit movement to slow the circulation of venom.

Seek Medical Attention Immediately:

Regardless of the type of snake or whether you think it's venomous, seek medical attention without delay. Emergency medical services should be contacted, and the victim should be transported to the nearest medical facility.

Immobilize the Bite Area:

Immobilize the bitten limb or body part to slow the spread of venom. Use a splint or sling to keep the affected area as still as possible. Avoid using a tourniquet, as it can cause more harm than good.

Keep the Bite Below Heart Level:

If possible, keep the bitten limb below the level of the heart. This can help slow the circulation of venom.

Remove Constrictive Items:

Remove tight clothing, jewelry, or accessories near the bite site, as swelling may occur. This can help prevent complications as the affected area swells.

Clean the Bite Area:

Wash the bite area gently with soap and water, but do not use ice or a very cold compress. Avoid making incisions at the bite site, as this can lead to additional complications.

Do Not Suck or Cut the Wound:

Traditional methods like sucking the venom out or cutting the wound are not recommended. These actions can cause more harm and are not effective in preventing the spread of venom.

Stay Hydrated:

Encourage the victim to stay hydrated by drinking water. However, alcohol and caffeine should be avoided, as they can speed up the body's absorption of venom.

Observe for Signs of Shock:

Watch for signs of shock, such as weakness, confusion, or a rapid pulse. If shock is suspected, keep the victim lying down with their legs elevated, unless this position causes pain.

Provide Information to Medical Professionals:

If possible, provide medical professionals with information about the snake, such as its color, size, and markings. However, do not attempt to capture or kill the snake, as this could lead to further injury.

Receive Antivenom Treatment:

In cases of venomous snake bites, the administration of antivenom is a critical component of treatment. Antivenom works to neutralize the effects of the snake venom.

Remember, time is of the essence when dealing with snake bites. Even if symptoms are initially mild, they can escalate rapidly. First aid is not a substitute for professional medical care. Snake bites, particularly those from venomous snakes, require immediate attention from healthcare professionals. The information provided here is intended as a general guideline, and it is crucial to follow specific recommendations from local health authorities and medical experts.

HARIET TUBMAN..........The Iron Lady!

Harriet Tubman, often referred to as "The Moses of Her People" or "The Conductor of the Underground Railroad," was a courageous African American abolitionist and political activist. Her life was marked by remarkable contributions to the abolition of slavery and her tireless efforts to aid enslaved individuals in escaping to freedom....The Iron Lady!

Birth and Childhood:

Harriet Tubman was born into slavery around 1820 or 1822 in Dorchester County, Maryland. Her birth name was Araminta Ross, and she later adopted the name Harriet Tubman upon marrying John Tubman.

Family Background:

Harriet Tubman was born into slavery, and her family experienced the harsh conditions of bondage. Her early years were marked by the brutality of slavery, and she suffered physical abuse.

Head Injury and Vision Loss:

During her childhood, Harriet Tubman sustained a head injury when she was hit by a heavy metal weight. This injury resulted in lifelong health issues, including seizures and periodic vision loss.

Escape from Slavery: In 1849, Harriet Tubman escaped slavery, leaving her family behind. This marked the beginning of her remarkable career as a conductor on the Underground Railroad, a network of safe houses and secret routes that helped enslaved individuals reach freedom in the Northern states or Canada.

Underground Railroad Leadership:

Tubman made numerous perilous journeys back into slaveholding states, guiding approximately 70 enslaved individuals to freedom. Her courage and resourcefulness earned her the nickname "Moses" among those she helped escape. During the time when Harriet Tubman was an active conductor on the Underground Railroad and a fugitive slave herself, there were indeed rewards offered for her capture.

Fugitive Slave Act:

The Fugitive Slave Act of 1850, part of the Compromise of 1850, was a federal law that required the capture and return of escaped slaves to their owners, even if they had reached free states or territories. The act imposed severe penalties on those who assisted escaped slaves, making it a federal offense to harbor or aid fugitive slaves.

Rewards for Harriet Tubman:

As Tubman became more well-known for her activities in helping enslaved individuals escape, particularly her work on the Underground Railroad, rewards were offered for her capture.

Slaveholders and those sympathetic to the pro-slavery cause sought to apprehend Tubman and put an end to her efforts.

Despite the risks and the rewards offered for her capture, Harriet Tubman continued her work as a conductor on the Underground Railroad. Her determination and fearlessness were evident in her multiple journeys back into slaveholding states to guide others to freedom. Her success in leading dozens of enslaved individuals to liberty earned her the nickname "Moses." Tubman became skilled at evading capture, utilizing different routes and methods to navigate through the Underground Railroad network. She often traveled at night, relying on the North Star and her knowledge of the landscape to avoid detection.

Civil War Service:

During the American Civil War, Tubman worked for the Union Army as a nurse, cook, and spy. She played a crucial role in the Combahee River Raid, a military operation that liberated more than 700 enslaved people.

Advocacy for Women's Suffrage:

After the Civil War, Tubman continued her activism by advocating for women's suffrage. She worked alongside suffragists like Susan B. Anthony and Elizabeth Cady Stanton.

Marital Life:

In 1844, Harriet Tubman married John Tubman, a free Black man. Unfortunately, John Tubman did not share Harriet's desire for freedom, and after her escape, she was unable to convince him to accompany her. In 1869, she married Nelson Davis, a Civil War veteran. The couple adopted a daughter named Gertie.

Later Life and Death:

Harriet Tubman spent her later years advocating for various causes, including women's rights and the care of the elderly. She used her own money to purchase property in Auburn, New York, where she established a home for elderly and indigent African Americans.

Death:

Harriet Tubman passed away on March 10, 1913, surrounded by friends and family. She was buried with military honors at Fort Hill Cemetery in Auburn, New York.

Harriet Tubman's legacy is one of resilience, bravery, and unwavering commitment to justice. Her contributions to the abolitionist movement, her efforts on the Underground Railroad, and her advocacy for the rights of marginalized communities continue to inspire generations. Harriet Tubman remains an iconic figure in American history, recognized for her pivotal role in the fight against slavery and her dedication to the pursuit of freedom and equality.

NELSON MANDELA..........Vs. the Apartheid Government!

Early Life:

Nelson Rolihlahla Mandela was born on July 18, 1918, in the village of Mvezo in Umtata, then part of South Africa's Cape Province. He belonged to the Thembu royal family and received the forename "Rolihlahla," meaning "pulling the branch of a tree" or metaphorically, "troublemaker." He was given the English forename "Nelson" by a teacher, a common practice at the time, and it became the name by which he would be universally known. Mandela's father died when he was a young boy, and he was adopted by Chief Jongintaba Dalindyebo, the regent of the Thembu people.

This association exposed Mandela to the politics, leadership, and legal proceedings within the Thembu royal house, influencing his later choices in life. Mandela grew up in the rural village of Qunu after moving from Mvezo. His childhood was shaped by the traditions, customs, and values of the Thembu community. He underwent the rites of passage into manhood, which were significant cultural events in Thembu tradition.

Education and Law Career:

Mandela's guardians ensured he received an education, and he attended a local mission school. This early exposure to formal education laid the foundation for his later studies in law and his engagement in political activism.

Mandela pursued his education at the University of Fort Hare and later at the University of Witwatersrand, where he studied law.

In 1942, he joined the African National Congress (ANC), becoming actively involved in anti-apartheid activism.

The Apartheid system:

The apartheid system was a legalized system of racial segregation and discrimination that existed in South Africa from 1948 to the early 1990s.

The apartheid government, led by the National Party, implemented a series of laws that institutionalized racial discrimination. The Population Registration Act of 1950 classified

South African residents by race, designating people as white, colored, Indian, or black. The Group Areas Act of 1950 enforced residential segregation, assigning specific areas for each racial group.

The government introduced pass laws that restricted the movement of black South Africans. Non-white individuals were required to carry passes, known as "dompas," which specified where they were allowed to live, work, and travel. Failure to produce a valid pass could result in arrest and imprisonment.

The Bantu Education Act of 1953 established a separate and inferior education system for black South Africans. The government aimed to prepare black students for menial jobs and prevent them from receiving an education that would challenge the existing social order. Employment opportunities were also limited by discriminatory labor laws. Public facilities, including schools, hospitals, beaches, and transportation, were segregated. Non-white individuals were forced to use separate facilities that were often of lower quality compared to those reserved for whites.

ANC Activism:

Mandela's early activism was nonviolent, focusing on legal challenges and protests against apartheid policies.

He was a key figure in the formation of the ANC Youth League in 1944, advocating for a more radical approach to ending apartheid.

Defiance Campaign and Apartheid Resistance:

In 1952, Mandela played a crucial role in organizing the Defiance Campaign against unjust apartheid laws, advocating nonviolent resistance.

The campaign marked a shift towards more assertive activism, setting the stage for increased resistance against apartheid.

Armed Struggle - Umkhonto we Sizwe:

Frustrated by the government's crackdown on peaceful protests, Mandela, along with others, formed the armed wing of the ANC, Umkhonto we Sizwe, in 1961. The decision to resort to armed struggle was driven by the realization that peaceful methods of protest were being met with increasing violence and repression by the apartheid government. The Sharpeville Massacre in 1960, where police opened fire on peaceful protesters, killing 69 people, highlighted the brutality of the regime. The ANC leadership, including Mandela, concluded that peaceful resistance alone might not be sufficient to bring about meaningful change.

Umkhonto we Sizwe, which translates to "Spear of the Nation" in Zulu, was established with the aim of conducting acts of sabotage against government installations, transportation networks, and other symbols of apartheid. Nelson Mandela played a key role in the formation of Umkhonto we Sizwe and became its first commander. In the early days, the organization was relatively small and lacked resources. Mandela, along with other leaders, engaged in fundraising and sought assistance from sympathetic countries to support the armed struggle. The organization engaged in acts of sabotage against government installations and infrastructure.

Arrest and Imprisonment:

Mandela was arrested in 1962 and sentenced to five years in prison for incitement and leaving the country without permission.

In 1964, during the Rivonia Trial, Mandela and several co-accused were sentenced to life imprisonment for plotting to overthrow the government.

Mandela spent 18 of his 27 years in prison on Robben Island, enduring harsh conditions and limited communication with the outside world.

His imprisonment turned him into a global symbol of the anti-apartheid struggle, garnering international support for his release.

Negotiations and Release:

In 1990, President F.W. de Klerk announced Mandela's release, signaling a commitment to ending apartheid.

De Klerk's announcement was met with both national and international attention. Mandela's release was not only a personal victory for him but also a symbolic victory for the anti-apartheid movement and the international community that had been pressuring South Africa to end its discriminatory policies.

Mandela emerged from prison as a revered leader, committed to reconciliation and negotiations for a democratic South Africa.

Transition to Democracy:

Mandela played a central role in the negotiations to end apartheid and establish a democratic, non-racial government. President F.W. de Klerk and Nelson Mandela jointly received the Nobel Peace Prize in 1993 for their efforts in dismantling apartheid and establishing a democratic South Africa. The release of Mandela in 1990 marked a momentous step towards the transformation of South Africa into a more just and inclusive society, and it remains a symbol of hope and reconciliation.

In 1994, South Africa held its first democratic elections, and Mandela became the country's first black president.

Presidency and Reconciliation:

Mandela's presidency (1994-1999) focused on reconciliation between South Africa's racially divided communities.

He established the Truth and Reconciliation Commission to address past human rights abuses and promote healing.

Personal life:

Mandela was married three times. His first marriage was to Evelyn Ntoko Mase in 1944, and the couple had four children. They divorced in 1957. His second wife, Winnie Madikizela-Mandela, whom he married in 1958, played a significant role in the anti-apartheid struggle but the marriage ended in divorce in 1996. Mandela's third marriage was to Graça Machel, the widow of former Mozambican President Samora Machel. They married in 1998.

Retirement and Legacy:

After serving one term as president, Mandela retired from politics but remained an influential figure in global affairs.

He dedicated his post-presidential years to charitable work, advocating for peace, reconciliation, and HIV/AIDS awareness.

Death:

Nelson Mandela passed away on December 5, 2013, leaving behind a legacy of courage, forgiveness, and the triumph of justice over oppression.Nelson Mandela's life journey is a testament to the power of resilience, forgiveness, and the unwavering commitment to justice. His legacy continues to inspire people around the world.

THE DAY OF THE DEAD..........Dia de los Muertos!

Dia de los Muertos, or the Day of the Dead, is a vibrant and deeply rooted Mexican tradition that celebrates and honors deceased loved ones. This multi-day festival, which has indigenous roots and is influenced by Catholicism, typically takes place from October 31st to November 2nd. While it coincides with the Catholic holidays of All Saints' Day and All Souls' Day, Dia de los Muertos has its unique customs and symbols.

Pre-Columbian Roots:

The celebration of death in Mesoamerican cultures, such as the Aztecs, is well documented. These civilizations believed in an afterlife and considered death as a natural part of the human cycle. Rituals and ceremonies honoring the deceased were common, often involving offerings of food, personal belongings, and even human sacrifices.

Aztec Festival of Mictecacihuatl:

The Aztecs dedicated a month-long celebration to Mictecacihuatl, the Lady of the Dead, corresponding to the modern calendar months of August and September. The festivities included rituals, dances, and offerings to honor the departed.

Spanish Influence:

When the Spanish conquistadors arrived in the Americas in the 16th century, they brought with them Catholicism. The Catholic Church sought to integrate indigenous traditions with Christian beliefs, and as a result, some of the indigenous rituals were merged with Catholic holidays such as All Saints' Day (November 1st) and All Souls' Day (November 2nd).

All Saints' Day and All Souls' Day:

The Catholic celebrations of All Saints' Day and All Souls' Day, dedicated to remembering and praying for the deceased, coincided with existing indigenous practices. Over time, the indigenous traditions and the Catholic holidays became intertwined, giving rise to the modern Dia de los Muertos.

Blend of Indigenous and Catholic Beliefs:

Dia de los Muertos retained many elements of its indigenous roots, including the building of altars, the use of marigolds, and the belief that the spirits of the deceased return to the world of the living during this time. The concept of death is often approached with humor and celebration rather than somberness.

Evolution of Traditions:

Over the centuries, Dia de los Muertos has evolved and adapted to various regional customs and cultural influences within Mexico. Different regions have unique ways of celebrating, with variations in the types of offerings, foods, and rituals.

Contemporary Celebration:

In the 20th and 21st centuries, Dia de los Muertos has gained international recognition, with celebrations extending beyond Mexico's borders. The iconic imagery of sugar skulls, marigolds, and Catrina figures has become synonymous with the festival.

Altars (Ofrendas):

Altars are a central element of Dia de los Muertos. Families create ofrendas, or offerings, on these altars to honor and remember their departed loved ones. These altars are adorned with photographs of the deceased, along with candles, flowers (especially marigolds), papel picado (colorful paper cutouts), and the favorite foods and beverages of the departed. The idea is to create a welcoming space for the spirits to return and enjoy the offerings.

Cemetery Visits:

Families often visit cemeteries during Dia de los Muertos to clean and decorate the gravesites of their loved ones. It is a communal and social occasion where families come together to remember and celebrate the lives of those who have passed away. Grave sites are often adorned with candles, flowers, and other offerings.

Sugar Skulls (Calaveras de Azúcar):

Sugar skulls are colorful and intricately decorated candies made from sugar, and they often feature the names of the deceased. These skulls are both a representation of death and a symbol of the sweetness of life. They are commonly used as offerings on altars and as gifts.

Pan de Muerto (Bread of the Dead):

Pan de Muerto is a special bread made during Dia de los Muertos. It is a sweet, often braided bread with bone-shaped decorations on top, symbolizing the bones of the deceased. Families place this bread on altars as an offering.

Calacas and Catrinas:

Calacas are colorful and whimsical skeleton figures, and Catrinas are elegant, dressed-up female skeletons. These figures are prevalent during Dia de los Muertos and are used in various forms of art, including sculptures, paintings, and costumes. They serve as a reminder of the inevitability of death and are often portrayed engaging in everyday activities.

Candlelight Vigils and Processions:

Many communities organize candlelight vigils and processions during Dia de los Muertos. Participants carry candles and sometimes dress in elaborate costumes and face paint. These events are a way for the community to come together and remember those who have passed away.

Symbolism of Marigolds:

Marigolds, or cempasúchil, are considered the flower of the dead and are believed to help guide the spirits back to the world of the living. They are often used to decorate altars and gravesites.

Dia de los Muertos is a unique and beautiful expression of Mexican culture that emphasizes the continuity of life and the importance of remembering and honoring those who have come before. It's a celebration that combines joyful remembrance with cultural and artistic expressions, providing a meaningful way for communities to connect with their heritage and celebrate the cycle of life and death.

THE AUTOBAHN..........Fast and Furious!

The Autobahn, often referred to as the "fast and furious" highway system, is a network of high-speed expressways in Germany known for its unique feature of having sections with no specific speed limits. Here are some key points to discuss about the Autobahn:

History:

The concept of the Autobahn was first developed in the 1920s, and the first section opened in 1932. However, it wasn't until the 1950s and 1960s that the network expanded significantly, becoming a crucial part of Germany's transportation infrastructure.

Infrastructure:

The Autobahn is characterized by its well-engineered roadways, designed to handle high-speed traffic. The highways have multiple lanes, advanced signage, and well-maintained surfaces to ensure safety and efficiency.

No Specific Speed Limits!

One of the most famous aspects of the Autobahn is that, unlike many highways around the world, it has sections with no specific speed limits. While there are recommended speed limits (Richtgeschwindigkeit) of 130 km/h (about 81 mph), drivers can legally go faster if road conditions and their vehicle's capabilities allow.

Speed Limits in Construction Zones:

It's important to note that speed limits are enforced in construction zones and certain urban areas. Additionally, during adverse weather conditions, temporary speed limits may be imposed for safety reasons.

Driver Responsibility:

The lack of a specific speed limit places a significant emphasis on driver responsibility and adherence to traffic rules. Drivers are expected to exercise judgment, taking into consideration their vehicle's capabilities, traffic conditions, and the safety of themselves and others.

High Performance Vehicles:

The Autobahn is often associated with high-performance and luxury vehicles capable of reaching and maintaining high speeds. Manufacturers, particularly German ones, often test and showcase their vehicles' performance on the unrestricted sections.

Safety Record:

Despite the absence of specific speed limits on some sections, the Autobahn has a relatively good safety record. German authorities attribute this to factors such as rigorous driver training, strict vehicle maintenance standards, and a strong emphasis on responsible driving.

Tourist Attraction:

The Autobahn has become a tourist attraction for car enthusiasts worldwide who want to experience driving on a high-speed highway with no specific speed limits. Car rental agencies often offer high-performance vehicles to those seeking the thrill of driving on the Autobahn.

Environmental Concerns:

The high speeds and frequent use of high-performance vehicles on the Autobahn have raised environmental concerns, particularly in terms of air pollution and carbon emissions. Efforts have been made to address these issues and promote sustainable transportation.

The Autobahn is a symbol of Germany's engineering prowess and is famous for its sections with no specific speed limits. It reflects a unique approach to highway design and traffic management, emphasizing responsible driving and the importance of well-maintained infrastructure. The Autobahn remains an iconic element of German culture and is a must-experience for automotive enthusiasts.

FLAMENCO..........Dancing with style!

Flamenco is a passionate and highly expressive art form that encompasses music, singing, dance, and even handclaps and finger snaps. Originating from the Andalusian region in southern Spain, Flamenco is deeply rooted in the cultural heritage of the Spanish Romani, or Gypsy, communities, as well as elements from Moorish, Jewish, and Andalusian folk traditions. At its core, Flamenco is a powerful means of emotional expression, allowing performers to convey a wide range of feelings through movement, rhythm, and music.

History:

The history of Flamenco is a rich tapestry that weaves together diverse cultural influences, historical events, and the artistic expressions of the Spanish Romani, or Gypsy, communities in the Andalusian region of southern Spain. While the exact origins of Flamenco are difficult to trace, its development can be understood through various historical and cultural contexts:

Gypsy Roots (15th-18th centuries):

The Gypsies arrived in Spain around the 15th century, and their music and dance traditions intertwined with the existing musical forms of Andalusia, including Moorish, Jewish, and Andalusian folk elements. This fusion laid the groundwork for what would later become Flamenco.

Early Forms and Influences (18th century):

During the 18th century, Flamenco began to take shape in the Andalusian region, particularly in the communities of Triana (Seville) and Jerez de la Frontera. It incorporated elements of Andalusian folk music, Arabic scales, and the rhythmic structures of Moorish and Gypsy traditions.

Café Cantantes and Popularization (19th century):

In the 19th century, Flamenco gained popularity in the urban centers of Andalusia, particularly through the emergence of "Café Cantantes" – cafes where Flamenco performances became a regular feature. This era saw the rise of professional Flamenco artists, including singers, dancers, and guitarists.

Golden Age (Late 19th to Early 20th centuries):

The late 19th and early 20th centuries are considered the "Golden Age" of Flamenco. During this period, Flamenco evolved into a more structured and codified art form. Notable artists like Antonio Chacón and Manuel Torre played crucial roles in shaping Flamenco's distinctive styles.

Flamenco Styles (Palos):

Over time, distinct styles within Flamenco, known as "palos," emerged. These include the solemn "Soleá," the festive "Alegrias," the rhythmic "Bulerías," and many others. Each palo has its own characteristic rhythm, melody, and thematic elements, contributing to the diversity of Flamenco.

International Influence (20th century):

In the 20th century, Flamenco expanded its reach beyond Spain and gained international recognition. Flamenco artists started touring globally, and the art form influenced various genres, including jazz and world music.

Flamenco in Film and Popular Culture:

Flamenco found its way into mainstream culture through films and popular music. The art form's dramatic and emotive qualities made it a captivating subject for filmmakers, while its rhythms and melodies influenced a wide range of musical genres.

Contemporary Flamenco (21st century):

In the 21st century, Flamenco continues to evolve and adapt. Modern Flamenco artists often experiment with fusion, incorporating elements of jazz, rock, and other genres while still preserving the core traditions. Flamenco festivals, schools, and events around the world contribute to the ongoing global appreciation of this art form.

Dancing with Style!

Intense Emotion and Expression:

Flamenco dance is renowned for its intense emotional expression. Dancers, known as "bailaores" (male) or "bailaoras" (female), use intricate footwork, intricate hand and arm movements, and facial expressions to convey a wide spectrum of emotions – from joy and passion to sorrow and defiance.

Posture and Elegance:

Flamenco dance is characterized by a distinctive posture, where the upper body remains upright and the arms and hands are elegantly positioned. This upright stance allows for a powerful connection between the dancer and the audience, emphasizing the emotional intensity of the performance.

Compás and Rhythm:

Rhythm, known as "compás," is a fundamental element of Flamenco. Dancers synchronize their movements with the complex rhythms created by the guitar, percussion, and singing. The intricate footwork, or "zapateado," is a rhythmic showcase that adds a percussive element to the performance.

Costumes and Accessories:

Flamenco dancers wear vibrant and elaborate costumes that enhance the visual impact of their movements. Traditional costumes often include colorful dresses with layers of ruffles, fringes, and lace for female dancers, while male dancers may wear tailored suits or traditional Andalusian attire. Accessories like fans and shawls are often incorporated into the choreography, adding flair to the performance.

Improvisation and Interaction:

Flamenco is known for its improvisational nature, allowing dancers the freedom to express their emotions in the moment. Performances often involve interaction between dancers, musicians, and singers, creating a dynamic and collaborative atmosphere on stage.

Cultural Influence and Global Appeal:

Over the years, Flamenco has transcended its cultural origins and gained international acclaim. It has inspired artists and dance enthusiasts worldwide, leading to the formation of Flamenco schools and festivals in various countries. The fusion of Flamenco with other dance forms and musical genres has resulted in innovative and contemporary expressions of this traditional art.

Flamenco dance is a captivating and dynamic form of artistic expression, representing the soulful spirit of Andalusian culture and captivating audiences with its emotional intensity and stylistic richness.

THE MAGNA CARTA..........Of Kings and Queens!

The Magna Carta, Latin for "Great Charter," is a historical document with profound significance in the development of constitutional governance and the rule of law. It was sealed on June 15, 1215, by King John of England at Runnymede, a meadow along the River Thames. The Magna Carta is often considered a foundational text in the establishment of principles that limit the power of the monarchy and protect the rights of subjects.

King John's Reign:

King John faced numerous challenges during his reign, including conflicts with the barons, financial difficulties, and military setbacks. He was the youngest son of King Henry II of England and Eleanor of Aquitaine. The discontent among the barons and the desire for greater protection of their rights and privileges led to the drafting of the Magna Carta.

Conflicts with Barons:

King John's reign was characterized by strained relationships with the barons, the powerful landowning nobility of medieval England. John's heavy taxation, arbitrary decisions, and perceived abuse of feudal rights led to deep-seated resentment among the barons. The king's demanding financial policies, including onerous taxes to fund military campaigns, exacerbated tensions.

Financial Difficulties:

King John faced significant financial challenges during his rule. His military endeavors, particularly the costly efforts to regain lost territories in France, strained the royal treasury. The king's ambitious military campaigns often ended in failure, draining resources and exacerbating his financial woes. To fund these endeavors, he imposed heavy taxes on his subjects, further intensifying discontent.

Military Setbacks:

King John's reign witnessed a series of military setbacks, most notably the loss of key territories in France. The defeat in the Anglo-French War and the loss of Normandy to the French King Philip II in 1204 were particularly humiliating. These military failures not only weakened John's political standing but also fueled discontent among the barons, who questioned the king's ability to protect and govern the realm effectively.

Abuse of Royal Powers:

King John was notorious for his arbitrary and often tyrannical exercise of royal power. His disregard for traditional feudal customs and his penchant for favoritism further alienated the barons. Additionally, John's treatment of widows, his exploitation of wardships, and the imposition of arbitrary fines created an atmosphere of mistrust and resentment.

Legal and Feudal Grievances:

The barons were particularly concerned about the violation of their legal and feudal rights. Arbitrary arrests, unfair trials, and the manipulation of inheritance practices by the king fueled grievances. The barons sought a restoration of the traditional rights and privileges they believed were guaranteed by customary law and the feudal system.

Role of Archbishop Stephen Langton:

Archbishop Stephen Langton played a crucial role in mediating between King John and the discontented barons. Langton, appointed by the Pope, sided with the barons in their efforts to curb the king's powers. His involvement and diplomatic skills contributed to the eventual negotiations that led to the Magna Carta.

Sealing of the Magna Carta (1215):

In the face of mounting pressure, King John reluctantly agreed to negotiate with the barons. The negotiations took place at Runnymede, a meadow along the River Thames, and resulted in the sealing of the Magna Carta on June 15, 1215. The document outlined a series of provisions that aimed to address the barons' grievances and establish principles of good governance.

While the Magna Carta was initially a peace treaty between King John and the barons, it went on to acquire broader significance as a foundational document in the development of constitutional principles, limiting the power of the monarchy and laying the groundwork for the rule of law in England and beyond.

Key Provisions and Themes:

Rule of Law:

The Magna Carta established the principle that no one, including the king, is above the law. It emphasized the idea that justice should be administered according to established legal procedures and not subject to arbitrary decisions by the ruler.

Protection of Barons' Rights:

A significant portion of the Magna Carta focused on the rights of the barons, the noble landowners of medieval England. It addressed concerns related to taxation, inheritance, and the king's treatment of barons, ensuring that the monarch could not act without constraint.

Due Process and Habeas Corpus:

The Magna Carta introduced the concepts of due process and habeas corpus. It emphasized that no free man could be imprisoned, outlawed, or deprived of property without a lawful judgment by his peers or the law of the land.

Consent and Consultation:

The document asserted the principle that major decisions, especially those related to taxation, required the "common counsel" or consent of the realm. This laid the groundwork for the development of representative government and parliamentary authority.

Limitations on Royal Power:

The Magna Carta sought to limit the arbitrary exercise of royal power. It established guidelines for the behavior of the king and his officials, ensuring that justice and governance were conducted within defined legal frameworks.

Symbol of Liberty:

Over time, the Magna Carta became a symbol of liberty and the protection of individual rights. It inspired subsequent generations in England and beyond to champion the idea that rulers should be accountable to the law.

Legal Foundations:

The Magna Carta laid the groundwork for constitutional principles that influenced the development of the English legal system. Its ideas are often cited in the evolution of legal thought and the drafting of subsequent legal documents.

Influence on Constitutionalism:

The Magna Carta is considered a precursor to modern constitutionalism. Initially a specific agreement between King John and his barons, its enduring legacy lies in the broader principles it articulated, which continue to shape notions of justice, governance, and individual rights in the modern world.

THE BLUE MOSQUE.........Created and Constructed!

The Blue Mosque, officially known as the Sultan Ahmed Mosque, is one of the most iconic and historically significant landmarks in Istanbul, Turkey. Often referred to as the "Blue Mosque" due to the striking blue tiles that adorn its interior, it stands as a masterpiece of Ottoman architecture and a symbol of the grandeur of the Ottoman Empire. The mosque is a testament to the skill and craftsmanship of the architects and artisans who contributed to its construction.

Architectural Brilliance:

The Blue Mosque was commissioned by Sultan Ahmed I and constructed between 1609 and 1616.

The chief architect of the mosque was Sedefkâr Mehmed Ağa, a renowned architect of the Ottoman Empire.

Inspiration from Hagia Sophia:

The Sultan Ahmed Mosque was designed to rival the nearby Hagia Sophia, a former Byzantine cathedral that had been converted into a mosque after the Ottoman conquest of Constantinople in 1453. The architects drew inspiration from the Hagia Sophia's architectural elements while incorporating new innovations.

Materials and Construction Techniques:

The mosque was constructed using a mix of traditional Ottoman and Byzantine architectural styles. The exterior features a cascade of domes and six slender minarets, a distinctive feature that caused controversy at the time, as it equaled the number of minarets of the Grand Mosque in Mecca.

Interior Decoration:

The interior of the Blue Mosque is known for its exquisite decoration, particularly the blue tiles that cover the walls. The tiles were produced in the city of Iznik and are a defining feature that gives the mosque its nickname. The intricate patterns, calligraphy, and floral designs create a visually stunning and serene atmosphere.

Courtyard and Fountain:

The mosque is surrounded by a large courtyard with a central fountain, providing a tranquil space for worshippers. The courtyard is enclosed by a colonnade, and the main entrance is through a monumental gateway.

Prayer Hall:

The prayer hall of the Blue Mosque is spacious and grand, with a high central dome flanked by smaller domes and semi-domes. The interior is illuminated by numerous windows, creating a sense of openness and light.

Religious Importance:

The Blue Mosque is an active mosque and serves as a place of worship for Muslims. Its location in the heart of Istanbul and its proximity to other historic sites make it a significant religious and cultural center.

Architectural Influence:

The architectural style of the Blue Mosque has had a lasting influence on Ottoman mosque design. Its elements, such as the cascading domes and the use of tiles, are reflected in subsequent mosque constructions in the Ottoman Empire.

Tourist Attraction:

Beyond its religious significance, the Blue Mosque is a major tourist attraction, drawing visitors from around the world who marvel at its architectural beauty and historical importance.

Cultural Symbol:

The Blue Mosque stands as a cultural symbol of the Ottoman period, representing the artistic and architectural achievements of the empire. It is a symbol of the blending of different cultural and artistic traditions in the heart of Istanbul.

The Blue Mosque, with its stunning architecture, intricate decoration, and historical importance, is a masterpiece that reflects the cultural and artistic richness of the Ottoman Empire. It continues to be a cherished landmark in Istanbul, attracting both worshippers and admirers of architectural beauty.

HIEROGLYPHICS.........Of Ancient Egypt!

Hieroglyphics are a system of writing that was used by the ancient Egyptians. The term "hieroglyphics" comes from two Greek words: "hieros" meaning "sacred" or "divine," and "glyphein" meaning "to carve." Hieroglyphics were a formal writing system that served multiple purposes, including religious texts, monumental inscriptions, administrative documents, and more. Here are key aspects to consider when discussing hieroglyphics in the context of ancient Egypt:

Origins and Development:

Hieroglyphics emerged around 3300 BCE during the late Predynastic Period in ancient Egypt. The writing system evolved from earlier pictorial and symbolic systems of communication. Over centuries, hieroglyphics became more standardized and sophisticated.

Writing Materials:

Hieroglyphics were often inscribed or carved into various materials, including stone, wood, pottery, and metal. The choice of material depended on the purpose and permanence of the inscription. Temples and tombs, for instance, often featured elaborate hieroglyphic inscriptions carved into stone.

Elements of Hieroglyphs:

Hieroglyphs encompass a wide range of symbols, including pictorial representations, ideograms, and phonetic signs. The writing system is a combination of logographic and alphabetic elements. Some symbols represented specific words or concepts, while others conveyed sounds.

Hieratic and Demotic Scripts:

In addition to hieroglyphics, the ancient Egyptians developed simplified scripts for everyday use. Hieratic, a cursive script derived from hieroglyphics, was used for administrative and literary texts. Demotic, another cursive script, evolved later and was used for a broader range of documents.

Religious Significance:

Hieroglyphics held great religious significance in ancient Egypt. Many inscriptions on temple walls and tombs contained religious texts, hymns, and spells. The Rosetta Stone, discovered in 1799, played a crucial role in deciphering hieroglyphics as it contained the same text in three scripts: hieroglyphics, demotic, and Greek.

Decipherment:

For centuries, hieroglyphics remained a mystery to scholars. The decipherment of hieroglyphics is often credited to Jean-François Champollion, who, in 1822, successfully deciphered the Rosetta Stone. This breakthrough opened up the ability to translate and understand the vast corpus of ancient Egyptian texts.

Use in Art and Monuments:

Hieroglyphics were extensively used in monumental art, particularly in the decoration of temples, pyramids, and tombs. Scenes from religious texts, historical events, and the lives of the pharaohs were often depicted alongside hieroglyphic inscriptions.

Papyrus Scrolls and Literary Works:

Hieroglyphics were also employed in the creation of papyrus scrolls, where they were written in a linear fashion rather than the intricate pictorial arrangement seen in monumental inscriptions. Literary works, religious texts, and administrative documents were written on papyrus using hieroglyphics.

Decline and Legacy:

The use of hieroglyphics declined with the spread of Christianity and the gradual disappearance of the ancient Egyptian civilization. However, their legacy endures as a fascinating aspect of ancient Egyptian culture and a key to understanding the civilization's history, beliefs, and daily life.

Hieroglyphics remain a captivating subject of study and have contributed immensely to our understanding of ancient Egyptian civilization. The ability to decipher this intricate writing system has provided valuable insights into the religious, cultural, and administrative aspects of one of the world's oldest civilizations.

THE HAGIA SOPHIA..........A Landmark of Landmarks!

The Hagia Sophia, located in Istanbul, Turkey, is an architectural masterpiece and a landmark that has witnessed the ebb and flow of history for over 1,500 years. Originally built as a cathedral, it later served as an imperial mosque and is now a museum.

Byzantine Cathedral (537-1453):

The Hagia Sophia, whose name means "Holy Wisdom" in Greek, was constructed as a cathedral during the reign of Byzantine Emperor Justinian I. Completed in 537 CE, it served as the principal church of the Eastern Orthodox Church for nearly a millennium. Its massive dome, intricate mosaics, and innovative architectural features made it an engineering marvel of its time.

Architectural Marvel:

The Hagia Sophia's architectural significance lies in its innovative use of a massive dome, semi-domes, and pendentives to create a spacious and awe-inspiring interior. The dome, with a diameter of 31 meters, was the largest in the world for nearly a thousand years. The building's vast open space and the use of light created a sense of grandeur.

Ottoman Mosque (1453-1935):

Following the Ottoman conquest of Constantinople in 1453, the Hagia Sophia was converted into a mosque by Sultan Mehmed II. The minarets were added, and the interior was adorned with Islamic elements such as calligraphy and Islamic medallions. The Hagia Sophia served as a mosque for nearly 500 years.

Mosaics and Artwork:

The Hagia Sophia originally featured a rich collection of mosaics depicting Christian themes. When it became a mosque, many of these mosaics were covered or plastered over. In the 20th century, during its conversion into a museum, efforts were made to uncover and restore these mosaics, revealing the layered history of the structure.

Secularization and Museum (1935-2020):

In 1935, under the leadership of Mustafa Kemal Atatürk, the founder of the Republic of Turkey, the Hagia Sophia was secularized and transformed into a museum. This move was part of broader efforts to modernize Turkey and separate religion from the state. As a museum, the Hagia Sophia became a symbol of Turkey's secular identity.

UNESCO World Heritage Site:

The Hagia Sophia was designated a UNESCO World Heritage Site in 1985, recognizing its outstanding universal value as a cultural and historical monument. Its inscription on the list highlighted its significance in both Byzantine and Ottoman history.

Reconversion into a Mosque (2020):

In July 2020, the Hagia Sophia underwent a significant change when the Turkish government issued a decree to convert it back into a mosque.

This decision sparked international discussions and reactions due to the structure's historical and cultural significance.

Contemporary Significance:

The Hagia Sophia remains a symbol of religious and cultural coexistence, reflecting the diverse history of Istanbul. Its architectural and artistic elements showcase the synthesis of Byzantine and Ottoman influences, making it a unique and cherished heritage site.

Visitor Attraction:

Throughout its various incarnations, the Hagia Sophia has attracted millions of visitors annually. Its immense historical, cultural, and architectural importance draws tourists, scholars, and admirers from around the world.

The Hagia Sophia is not only a landmark of architectural and engineering brilliance but also a symbol of the historical and cultural transformations that have shaped Istanbul and the broader region over centuries. Its status as a museum, mosque, or both reflects the complex layers of history that define this iconic structure.

THE HONEY BADGER.......... An Apex Predator??

The honey badger, scientifically known as Mellivora capensis, is a small to medium-sized mammal native to Africa, Southwest Asia, and the Indian subcontinent. It is known for its tenacity, intelligence, and remarkable adaptability. But does the honey badger qualify as an apex predator? Well, here are some key characteristics and aspects of the "The Most Fearless Animal in the World"....

Physical Characteristics:

Honey badgers have a stocky, muscular build with a broad head, strong claws, and loose, tough skin that makes it difficult for predators to grasp them.

They have a distinctive black and white coat and a conspicuous white stripe on their back.

Diet and Foraging:

Honey badgers are opportunistic omnivores, feeding on a wide range of prey. Their diet includes small mammals, birds, reptiles, insects, fruits, and, as their name suggests, honey. They are also known for raiding beehives, enduring stings from bees in the process.

Fearless and Tenacious Behavior:

Honey badgers are renowned for their fearless and tenacious nature. Despite their relatively small size, they are known to confront and challenge much larger predators, including lions and hyenas. Their loose, tough skin allows them to twist and turn within the grasp of a predator, making it difficult for them to be captured.

Adaptability:

Honey badgers are highly adaptable to different environments, ranging from grasslands and savannas to forests and deserts. They are primarily terrestrial but are capable climbers and swimmers. Their adaptability allows them to thrive in a variety of ecosystems.

Nocturnal Behavior:

Honey badgers are primarily nocturnal, meaning they are most active during the night. This behavior helps them avoid the heat of the day and reduces competition for resources with other diurnal animals.

Conservation Status:

The conservation status of honey badgers varies depending on the region, but they are generally classified as a species of "Least Concern" by the International Union for Conservation of Nature (IUCN). However, localized threats, such as habitat loss and persecution, can impact specific populations.

An Apex Predator?

Apex predators are typically at the top of the food chain and have few natural predators themselves. Honey badgers, while displaying formidable defensive capabilities, are not at the top of their respective ecosystems. They face predation from larger carnivores, such as big cats and hyenas.

Honey badgers however, are known for their toughness, adaptability, and fearlessness, but sadly, this alone doesn't qualify them as apex predators. They nonetheless, possess an unusual resilience that portray them as a formidable threat even to predators themselves!

MICHELANGELO..........Gifted Hands!

Michelangelo Buonarroti, commonly known as Michelangelo, was a renowned Italian Renaissance artist whose incredible talent left an indelible mark on the world of art and culture. Born on March 6, 1475, in Caprese, Italy, Michelangelo's "gifted hands" created some of the most celebrated works of art in history.....A legend indeed!

Early Life and Education:

Michelangelo showed early artistic talent and, at the age of 13, entered the workshop of the painter Domenico Ghirlandaio. Later, he joined the Medici family, where he had the opportunity to study classical sculpture in their gardens and developed a deep appreciation for ancient Greek and Roman art.

Sculpture:

Michelangelo is perhaps most famous for his sculpture. One of his early masterpieces is the "Pieta," a sculpture of the Virgin Mary holding the body of Christ. Completed when he was only 24, it demonstrated his exceptional skill in rendering the human form.

The statue of David is another iconic sculpture by Michelangelo. Carved from a single block of marble, it is a representation of the biblical hero David before his battle with Goliath. The statue is a symbol of strength, beauty, and human potential.

Sistine Chapel Ceiling:

Commissioned by Pope Julius II, Michelangelo spent four years (1508-1512) painting the ceiling of the Sistine Chapel in the Vatican. The masterpiece includes the famous frescoes depicting scenes from Genesis, such as the Creation of Adam. The complexity and grandeur of the Sistine Chapel ceiling established Michelangelo as a preeminent artist of his time.

Architectural Achievements:

Michelangelo was not only a sculptor and painter but also an accomplished architect. He designed the dome of St. Peter's Basilica in Rome, a structure that became one of the most iconic symbols of the city. His architectural contributions blended classical elements with innovative designs.

Poetry and Literature:

In addition to his visual arts, Michelangelo was a poet. His sonnets and poems, often expressing spiritual and philosophical themes, provide insights into his inner thoughts and struggles. His literary works, like his visual creations, contributed to the cultural legacy of the Renaissance.

Last Judgment:

Another notable fresco by Michelangelo is the "Last Judgment," located on the altar wall of the Sistine Chapel. Painted later in his career (1536-1541), it depicts the second coming of Christ and the final judgment. The fresco is known for its powerful depiction of the human form and emotional intensity.

Marital Life:

Michelangelo never married and had no children. He chose a life dedicated to art and spirituality, which may have contributed to his decision to remain single. Throughout his life, he maintained a strong focus on his work, and his passion for art often overshadowed other aspects of his personal life. He however, maintained significant relationships with various influential figures of his time. He developed friendships with other renowned artists, such as Leonardo da Vinci and Raphael, although his relationships were sometimes marked by rivalry and competition. Michelangelo also had close connections with patrons and powerful individuals, including the Medici family and Pope Julius II, who commissioned several of his notable works. He was deep rooted Christian had a profound faith in God. His Catholic upbringing influenced many of his works, and he found inspiration in religious themes. This spiritual connection is evident in masterpieces like the Sistine Chapel ceiling and the Pietà. Michelangelo's personal convictions often guided his choices and interactions.

Death:

In the last years of his life, Michelangelo continued to work in Rome. He had a long and successful career, leaving behind a legacy of masterpieces in sculpture, painting, and architecture. Some of his notable late works include the completion of the dome of St. Peter's Basilica in Vatican City. As Michelangelo aged, he experienced various health issues. In the years leading up to his death, he suffered from a number of ailments, including arthritis and kidney problems. Despite his physical decline, he remained dedicated to his work, and his passion for art did not wane. He eventually withdrew from social life and became increasingly introspective. He was a private individual, and his focus on his work often led to a solitary lifestyle. His contemporaries described him as reserved and sometimes difficult to approach.

After a long battle with kidney problems, arthritis and other ailments, Michelangelo gave up the ghost on February 18, 1564, in Rome. His death marked the end of an era in Italian art. His body was initially interred at the Basilica di Santa Croce in Florence, in accordance with his wishes. However, his nephew, Leonardo Buonarroti, later transported Michelangelo's remains to the family chapel in the Basilica di Santa Croce.

Legacy and Influence:

Michelangelo's influence on art and culture extended far beyond his lifetime. His dedication to the perfection of form, his mastery of anatomy, and his ability to convey emotion through art left an enduring impact on subsequent generations of artists. His works became a touchstone for the Baroque and Mannerist movements that followed.

Michelangelo's "gifted hands" produced a body of work that continues to captivate and inspire art enthusiasts and scholars alike. His ability to bring out the beauty of the human form, his technical brilliance, and his contributions to various artistic disciplines make him one of the most celebrated figures in the history of art.

LEONARDO DA VINCI.......... The Polymath!

Leonardo da Vinci is widely regarded as one of the most versatile and brilliant individuals in history. Often referred to as a "polymath" or "Renaissance man," Leonardo excelled in various fields, including painting, sculpture, anatomy, engineering, architecture, mathematics, music, and more. His diverse talents and accomplishments have left an enduring impact on art, science, and humanity as a whole.

Early Life:

Leonardo da Vinci was born on April 15, 1452, in Vinci, a small town in Italy, which is now located in the region of Tuscany. He was the illegitimate son of Ser Piero da Vinci, a notary, and Caterina, a peasant woman. His parents never married, and his father later married another woman, while his mother married someone else as well. Leonardo received a basic education in reading, writing, and arithmetic. Recognizing his artistic talent, his father apprenticed him to the renowned artist Andrea del Verrocchio in Florence when he was around 14 years old. Verrocchio was a prominent painter, sculptor, and goldsmith, and under his guidance, Leonardo learned various artistic skills.

Artistic Mastery:

Leonardo is perhaps best known for his extraordinary skills as a painter. His masterpieces include "The Last Supper" and the iconic "Mona Lisa." His innovative techniques, such as sfumato (a technique of blending colors and tones), and his meticulous attention to detail set him apart as a groundbreaking artist of the Renaissance.

Scientific Inquiry:

Leonardo was a keen observer of the natural world. He made numerous detailed anatomical sketches, studying the human body to gain a deeper understanding of anatomy and physiology. His notebooks are filled with sketches of the human skeleton, muscles, organs, and even fetal development. Leonardo's scientific curiosity extended to the study of plants, animals, geology, and hydraulics.

Engineering and Inventions:

Leonardo's notebooks are filled with designs for various machines and inventions, showcasing his remarkable engineering mind. He envisioned flying machines, war machines, hydraulic devices, and even a concept for an early helicopter. While many of his inventions were not built during his lifetime, they demonstrate his forward-thinking and innovative approach to problem-solving.

Architecture:

Leonardo explored architectural concepts, creating designs for fortifications, bridges, and city planning. His architectural sketches and plans reveal his understanding of geometry, proportion, and spatial relationships.

Mathematics:

Leonardo had a strong grasp of mathematics, and his notebooks contain mathematical studies and geometric diagrams. He applied mathematical principles to his art, ensuring precision and balance in his compositions.

Musical Talent:

Leonardo was also known for his musical abilities. He played several musical instruments, and he is credited with designing an instrument called the viola organista, a combination of a keyboard and strings.

Writing and Notebooks:

Leonardo's notebooks are a treasure trove of ideas, observations, sketches, and concepts. They contain over 13,000 pages of his thoughts on art, science, anatomy, engineering, and more. His habit of recording everything he observed or conceived contributed significantly to our understanding of his genius.

Marital Life:

There is no historical evidence to suggest that Leonardo da Vinci ever married or had any children. He was known for his intensely private and secretive nature, and much of his personal life remains undocumented. Due to the lack of concrete information, various theories and speculations have emerged regarding Leonardo's personal life. Some suggest that he may have been celibate, while others propose that he had relationships that were kept private. However, these theories are largely speculative and lack solid historical evidence.

Death:

Life in France:

In the later years of his life, Leonardo left Italy and entered the service of King Francis I, who appreciated and admired the artist's talents. The French monarch provided Leonardo with a residence at the Château de Cloux (now known as Clos Lucé) near Amboise.

Health Issues:

Leonardo da Vinci faced health challenges in his later years. It is believed that he suffered from a stroke in the months leading up to his death. This may have been a contributing factor to his declining health.

The Death of Leonardo da Vinci:

Leonardo da Vinci died on May 2, 1519, at Clos Lucé. The exact cause of his death is not definitively known, but it is widely believed that he passed away peacefully in his sleep. Some historical accounts suggest that he died in the presence of King Francis I. Initially, Leonardo was buried in the Chapel of Saint-Florentin at the Château d'Amboise. However, the chapel was destroyed during the

French Revolution, and the exact location of his grave became uncertain.

While Leonardo da Vinci's physical presence may have ended in 1519, his ideas and creations continue to captivate and inspire people around the world. His death marked the conclusion of a life dedicated to the pursuit of knowledge and artistic excellence, leaving behind a legacy that transcends centuries. Leonardo da Vinci's ability to seamlessly integrate art and science exemplifies the Renaissance spirit of intellectual curiosity and a holistic approach to knowledge. His legacy continues to inspire and influence diverse fields, making him a true polymath and a timeless symbol of human potential and creativity.

THE MOON LANDING HOAX CONSPIRACY..........True or False?

The Moon landing hoax conspiracy theory suggests that the United States faked the Apollo moon landings in the 1960s and 1970s. While this theory has persisted for decades, it is overwhelmingly false. The Apollo moon landings, conducted by NASA, were real, and there is an extensive body of evidence supporting the authenticity of these missions. Here are some key points debunking the Moon landing hoax conspiracy......

Photographic Evidence:

Numerous photographs were taken during the Apollo missions, and these images provide clear visual evidence of the astronauts on the Moon's surface. The lighting, shadows, and reflections captured in the photos are consistent with the unique conditions of the lunar environment.

Rock Samples:

The Apollo missions returned with rock samples from the Moon, and these have been extensively studied by scientists around the world. The composition of these lunar rocks is distinct from Earth rocks, providing tangible proof of the missions' extraterrestrial origin.

Laser Reflectors:

During the Apollo 11, 14, and 15 missions, astronauts placed retro reflectors on the Moon's surface. These devices reflect laser beams sent from Earth, and their continued use over the years provides ongoing evidence of human presence on the Moon.

Telemetry and Communication:

The communication signals between the astronauts and mission control were tracked by multiple ground stations around the world. These signals, including voice transmissions, telemetry data, and video feeds, were consistent with the characteristics of space communication and could not have been easily replicated on Earth.

Independent Verification:

The Apollo missions were tracked by independent organizations and countries, including the Soviet Union, which was in the midst of the Space Race with the United States. If the Moon landings had been faked, it is highly likely that the Soviet Union would have exposed the hoax.

Eyewitness Testimony:

Thousands of people were involved in the planning, execution, and support of the Apollo missions. Astronauts, scientists, engineers, and technicians provided consistent and credible eyewitness testimony about their roles in the missions.

Technological Limitations at the Time:

The technology available in the 1960s and 1970s was not advanced enough to convincingly simulate a Moon landing, especially given the scrutiny and expertise of the scientific community and the public.

Despite the overwhelming evidence supporting the authenticity of the Apollo Moon landings, conspiracy theories persist. These theories often rely on misconceptions, misinterpretations of data, and a general distrust of official narratives. However, the scientific consensus and the extensive body of evidence strongly support the fact that humans did land on the Moon during the Apollo missions.

THE LOST CITY OF ATLANTIS..........The Strait of Gibraltar!

The Lost City of Atlantis is a legendary and mythical island first mentioned by the ancient Greek philosopher Plato in his dialogues "Timaeus" and "Critias." According to Plato, Atlantis was a powerful and advanced civilization that existed around 9,000 years before his own time (circa 11,000 years ago). While the exact location of Atlantis remains a subject of speculation and debate, some theories have suggested that the Strait of Gibraltar, the narrow passage connecting the Atlantic Ocean to the Mediterranean Sea, may be associated with the legendary city.

Here are some aspects to consider when discussing the connection between the Lost City of Atlantis and the Strait of Gibraltar:

Plato's Account:

Plato's descriptions of Atlantis are found in his dialogues "Timaeus" and "Critias." According to these accounts, Atlantis was a prosperous and technologically advanced civilization with a powerful navy and a highly organized society. It eventually sank into the ocean in a catastrophic event.

Geographical Clues:

Plato provided certain geographical details about Atlantis, including its size, location, and proximity to other lands. Some interpretations of these details have led to speculations that the location of Atlantis might be in the region of the Strait of Gibraltar.

Arguments Against the Connection:

Critics argue that the story of Atlantis is more likely an allegory or philosophical tale created by Plato to convey moral lessons rather than a historical account. Additionally, the timeline provided by Plato does not align with archaeological and historical evidence for advanced civilizations of that scale in the purported time frame.

Geological Considerations:

The idea of Atlantis being located in the Strait of Gibraltar faces challenges from a geological perspective. The theory that a large landmass suddenly sank into the ocean contradicts known geological processes. Moreover, the Strait of Gibraltar has been closely studied, and there is no evidence of a submerged ancient city in that area.

Destruction:

Atlantis, according to Plato, faced a catastrophic event that led to its sinking into the ocean. The cause of this destruction is not explicitly stated, but it is described as a divine punishment for the city's moral decay and arrogance.

Alternative Theories:

Numerous alternative theories propose different locations for Atlantis, ranging from the Mediterranean to Antarctica or the Caribbean.

However, none of these theories have gained widespread acceptance within the scientific and archaeological communities.

While the idea of Atlantis being located in the Strait of Gibraltar has been suggested, it lacks empirical evidence and faces significant challenges from both a historical and geological standpoint. The Lost City of Atlantis remains one of the enduring mysteries of ancient history and continues to capture the imagination of people around the world, inspiring various theories and interpretations.

EL CHUPACABRA..........A Living Disaster!

El Chupacabra, which translates to "The Goat Sucker" in Spanish, is a legendary creature in folklore, particularly in Latin American and Puerto Rican cultures. Descriptions of El Chupacabra vary, but it is often depicted as a creature that attacks and drinks the blood of livestock, especially goats. While the legend has gained popularity and sparked numerous stories and sightings, there is no scientific evidence supporting the existence of such a creature. Instead, El Chupacabra is widely considered to be a mythical creature or an example of urban folklore.

Origins of the Legend:

The legend of El Chupacabra first emerged in Puerto Rico in the mid-1990s. Reports of mysterious livestock deaths, where the animals were found drained of blood, led to the belief in a creature that was responsible for these attacks.

Physical Descriptions:

Descriptions of El Chupacabra vary, but common elements include a creature about the size of a small bear or a large dog, with reptilian or alien features. It is often said to have spikes or quills on its back and fangs for sucking blood.

Livestock Attacks:

The legend attributes livestock deaths, particularly goats and other small animals, to El Chupacabra. Victims are often found with puncture wounds and blood drained from their bodies. However, many instances of supposed Chupacabra attacks have been debunked as predation by known animals or natural causes.

Global Phenomenon:

The legend of El Chupacabra quickly spread beyond Puerto Rico and became part of the folklore in various Latin American countries and even in parts of the United States. Sightings and reports of Chupacabra-like creatures have been claimed in different regions.

Media and Popular Culture:

El Chupacabra has become a prominent figure in popular culture, appearing in books, movies, TV shows, and other forms of entertainment. Its portrayal ranges from a fearsome, blood-sucking creature to a more comical or cartoonish depiction.

Scientific Skepticism:

The scientific community dismisses the existence of El Chupacabra as a real, biological creature. Many reported Chupacabra attacks have been attributed to known predators, such as dogs, coyotes, or other wildlife. The phenomenon is often seen as a result of fear, superstition, or misidentification of natural animal behaviors.

Cultural Significance:

The legend of El Chupacabra reflects cultural anxieties and fears about the unknown. It has become a part of the cultural fabric in certain regions, influencing storytelling and local traditions.

The legend of El Chupacabra has become a fascinating and enduring part of folklore, but the fact remains that there is no credible evidence supporting the existence of a creature matching its descriptions. Instead, the phenomenon is rooted in cultural stories, misunderstandings, and the human tendency to attribute unexplained events to mythical or supernatural causes.

SHERLOCK HOLMES.........Reason against Treason!

Sherlock Holmes is a fictional detective created by Sir Arthur Conan Doyle. He first appeared in the novel "A Study in Scarlet," published in 1887, and went on to become one of the most iconic and enduring characters in detective fiction. Holmes is known for his brilliant deductive reasoning, keen observational skills, and logical approach to solving complex mysteries.

Deductive Reasoning:

Sherlock Holmes is renowned for his deductive reasoning skills. He often solves crimes by carefully observing details that others might overlook and then drawing logical conclusions from the evidence at hand. Holmes frequently relies on his powers of deduction to unravel mysteries. One notable example of Sherlock Holmes showcasing his deductive reasoning skills can be found in the story "The Adventure of the Blue Carbuncle," which is part of the collection of short stories titled "The Adventures of Sherlock Holmes."

"The Adventure of the Blue Carbuncle,"

In this particular case, Holmes is presented with a hat and a goose, both left behind by someone who had been involved in a scuffle. Dr. John Watson, Holmes's friend and companion, brings these items to Holmes for analysis.

The Found Items:

Watson discovers a hat and a goose in the street. The goose has been dropped and trampled, and the hat shows signs of a struggle.

Observations:

Holmes carefully examines the hat and observes certain details, such as the type of hat, the condition of the brim, and the presence of a unique type of mud on it.

Deductive Reasoning:

Holmes deduces several key points based on his observations:

The hat belongs to a middle-aged man who has done manual labor.

The man is fairly well-off but has fallen on hard times recently.

He has been careless in his dress but has tried to improve his appearance.

The mud on the hat suggests a specific location in London.

The Missing Gem:

Holmes also discovers a precious gem, a blue carbuncle, inside the goose's crop. He deduces that the gem was lost during a struggle, and the goose unwittingly ingested it.

Conclusion:

Through his deductive reasoning, Holmes pieces together the events: The man who lost the hat had a quarrel with another man, and during the scuffle, the blue carbuncle was dropped. The gem was then swallowed by the goose as it fed on the street.

Holmes's deductions lead him to the owner of the hat, who turns out to be an innocent man wrongly accused of the theft of the blue carbuncle. Holmes's ability to unravel the mystery and identify the true sequence of events showcases his exceptional deductive reasoning skills.

Observational Skills:

Holmes pays meticulous attention to his surroundings and the people involved in a case. His keen observational skills allow him to notice subtle details, such as a person's clothing, mannerisms, or the condition of their personal items. These observations contribute to his ability to piece together the puzzle of a crime.

Logical Thinking:

Holmes approaches investigations with a scientific and logical mindset. He values reason and evidence over emotional reactions, and he often criticizes flawed reasoning. His emphasis on logic and evidence has made him a symbol of rational thinking in detective fiction.

Famous Cases:

Holmes is known for solving a variety of complex cases, including "The Hound of the Baskervilles," "The Adventure of the Speckled Band," and "The Sign of the Four." Each case showcases his analytical prowess and commitment to solving mysteries through reason.

Partnership with Dr. John Watson:

Dr. John Watson, Holmes's loyal friend and companion, often serves as the narrator in the stories. Watson is both fascinated and sometimes mystified by Holmes's extraordinary ability to reason and deduce. The dynamic between Holmes and Watson is central to the success of the Sherlock Holmes stories.

Legacy:

Sherlock Holmes has left an indelible mark on detective fiction and popular culture. The character has been adapted into numerous films, TV series, and other media. Holmes's emphasis on reason and logic has influenced subsequent generations of fictional detectives and real-life investigators.

Sherlock Holmes's ability to "reason" set him apart as the quintessential detective in the world of fiction. The detective's commitment to reason and deduction aligns with the idea of using logical thinking to uncover the truth in the face of criminal activities.

PHILADELPHIA..........A vibrant City!

Philadelphia, often referred to as the "City of Brotherly Love," is a vibrant and historically rich city located in the state of Pennsylvania, United States. Here are several aspects that contribute to Philadelphia's vibrancy:

Historical Significance:

Philadelphia played a pivotal role in the early history of the United States. It is where the Declaration of Independence was adopted in 1776, and the United States Constitution was drafted in 1787. As such, Philadelphia is often considered the birthplace of the nation.

Historic Landmarks:

The city is home to numerous historic landmarks, including Independence Hall, the Liberty Bell, and the National Constitution Center. These sites attract millions of visitors each year, offering a glimpse into the nation's founding history.

Cultural Diversity:

Philadelphia is known for its cultural diversity, with a rich tapestry of communities and neighborhoods representing various ethnicities and cultures. This diversity is reflected in the city's culinary scene, festivals, and cultural events.

Arts and Culture:

The city has a thriving arts and culture scene. The Philadelphia Museum of Art, home to the iconic "Rocky Steps," is one of the largest and most prestigious art museums in the country. The city also boasts numerous theaters, galleries, and music venues.

Education and Institutions:

Philadelphia is home to several prestigious universities, including the University of Pennsylvania, Drexel University, and Temple University. These institutions contribute to the city's intellectual vibrancy and innovation.

Green Spaces:

Philadelphia features beautiful parks and green spaces, such as Fairmount Park, one of the largest urban park systems in the United States. Residents and visitors alike can enjoy outdoor activities and events in these scenic areas.

Sports Enthusiasm:

Philadelphia is known for its passionate sports culture. The city's sports teams, including the Philadelphia Eagles (NFL), Philadelphia Phillies (MLB), and Philadelphia 76ers (NBA), have a dedicated fan base, and attending a game is a lively and communal experience.

Culinary Scene:

The city has a dynamic food scene, ranging from historic markets like Reading Terminal Market to modern and diverse restaurants. Philadelphia is particularly famous for its cheesesteaks, soft pretzels, and other local delicacies.

Revitalized Neighborhoods:

In recent years, various neighborhoods in Philadelphia have experienced revitalization, with new businesses, shops, and restaurants contributing to the overall vibrancy. Areas like Fishtown, Northern Liberties, and University City are examples of thriving communities.

Community Engagement:

Philadelphia has a strong sense of community engagement, with various events, festivals, and initiatives that bring residents together. Community gardens, farmers' markets, and local events contribute to the city's sense of vibrancy and connection.

Philadelphia's vibrancy is shaped by its rich history, cultural diversity, commitment to education, arts and culture scene, green spaces, and a strong sense of community. Whether exploring its historic sites, enjoying cultural events, or experiencing its lively neighborhoods, Philadelphia offers a dynamic and engaging urban experience.

THE BERMUDA TRIANGLE..........Known to the Unknown!

The Bermuda Triangle, also known as the "Devil's Triangle," is a loosely defined region in the western part of the North Atlantic Ocean. It is infamous for the alleged mysterious disappearances of ships and aircraft. While the concept of the Bermuda Triangle has captured the public imagination for decades, it's essential to note that the scientific community largely dismisses the idea of any mysterious or paranormal activity in this region. The Bermuda Triangle is not officially recognized as a hazardous or anomalous area by maritime or aviation authorities.

Geographical Boundaries:

The Bermuda Triangle's boundaries are typically considered to be points in Miami, Bermuda, and Puerto Rico, forming a triangular area in the western part of the North Atlantic Ocean.

Mysterious Disappearances:

The Bermuda Triangle gained notoriety due to reports of ships and aircraft allegedly disappearing under mysterious circumstances while traversing the area. Some well-known incidents include the disappearance of Flight 19 in 1945 and the case of the USS Cyclops in 1918.

Flight 19 Disappearance!

The disappearance of Flight 19 is one of the most famous aviation mysteries in history. It occurred on December 5, 1945, during a routine training mission in the Bermuda Triangle, an area in the western part of the North Atlantic Ocean known for numerous aircraft and ship disappearances. Flight 19 was a squadron of five U.S. Navy TBM Avenger torpedo bombers.

The flight took off from Naval Air Station Fort Lauderdale in Florida for a navigation training exercise, known as "Navigation Problem Number One." The plan was to fly east from Florida, conduct bombing practice over Hens and Chickens Shoals, then turn north and finally west to return to the base. However, things went awry during the exercise.

Several factors contributed to the confusion and ultimate disappearance of Flight 19:

Weather Conditions:

 The weather on that day was challenging, with overcast skies and poor visibility. Pilots reported difficulties in determining their positions.

Compass Malfunction:

It was reported that at least one of the aircraft experienced compass malfunctions. This could have led to navigation errors, especially considering the reliance on compasses for navigation during that time.

Communication Issues:

The squadron leader, Lieutenant Charles Taylor, was having trouble communicating with his pilots due to radio malfunctions. This made coordination and navigation even more challenging.

As the flight progressed, communication became increasingly sporadic, and the pilots were unable to determine their location accurately. The last radio transmission from Flight 19 suggested that they were lost and running low on fuel.

A rescue mission was launched, involving a Martin PBM Mariner flying boat. Shockingly, the rescue plane, Flight 19's intended rescuer, also disappeared without a trace during the search. The disappearance of Flight 19 has given rise to various theories, including the paranormal and extraterrestrial explanations associated with the Bermuda Triangle. However, most experts attribute the incident to a combination of navigational errors, adverse weather conditions, and communication problems.

The exact fate of Flight 19 remains unknown, as no wreckage or remains were ever found. The mystery surrounding the disappearance of Flight 19 has fueled speculation and contributed to the enduring fascination with the Bermuda Triangle.

The USS Cyclops!

The USS Cyclops was a U.S. Navy collier ship that mysteriously disappeared during World War I. Its disappearance remains one of the greatest maritime mysteries in history. The USS Cyclops was a

Proteus-class collier, and it played a crucial role in fueling the British fleet during the war.

Background:

The USS Cyclops was commissioned in 1910 and primarily used for transporting coal. It had a displacement of over 19,000 tons, making it one of the largest fuel ships in the U.S. Navy at the time.

In 1918, during World War I, the Cyclops was assigned to the Naval Overseas Transportation Service and tasked with delivering coal to U.S. and Allied naval bases.

Last Known Movements:

The USS Cyclops departed from Rio de Janeiro on February 16, 1918, with a load of manganese ore.

After stopping in Bahia, Brazil, it continued on its journey toward Baltimore, Maryland.

The ship was last heard from on March 4, 1918, when it sent a routine radio message indicating that it was experiencing engine troubles but still making progress.

Disappearance:

The USS Cyclops never reached its destination, and no distress signals were ever received. It disappeared without a trace.

Despite an extensive search, no wreckage or debris from the ship was found. The disappearance was declared "without a trace," and all 306 crew members were presumed lost.

Theories and Speculations:

The exact cause of the USS Cyclops' disappearance remains unknown, leading to various theories and speculations. Some of the proposed explanations include:

Mechanical Failure: Some believe that the ship's engines or other mechanical systems failed, leading to its sinking.

Storm: It's possible that the ship encountered a severe storm that caused it to sink, but no specific evidence supporting this theory has been found.

Enemy Action: There were unsubstantiated rumors of German U-boats operating in the area, leading to speculation that the USS Cyclops might have been a target.

The USS Cyclops' disappearance is considered one of the greatest unsolved maritime mysteries. It has fueled speculation and conspiracy theories over the years, with the Bermuda Triangle often invoked due to its general vicinity to the area.

Despite numerous investigations and theories, the fate of the USS Cyclops and the reason for its disappearance remain unresolved. The incident highlights the challenges and dangers faced by maritime vessels during times of war and the mysteries that can surround their disappearances.

Scientific Skepticism:

The scientific community attributes the alleged disappearances to natural phenomena, human error, and sometimes sensationalism. There is no empirical evidence supporting the existence of any supernatural or unusual forces in the Bermuda Triangle.

Common Explanations:

Various explanations have been proposed for incidents in the Bermuda Triangle, including navigational errors, human factors, violent weather, and the presence of underwater features such as deep-sea trenches and methane hydrate deposits. These factors are present in other parts of the world and are not unique to the Bermuda Triangle.

Selective Reporting:

Critics argue that the Bermuda Triangle mystery is perpetuated by selective reporting, where incidents from the region are highlighted while similar incidents in other parts of the world are overlooked. This selective reporting contributes to the perception of the Bermuda Triangle as uniquely dangerous.

Official Stance:

Maritime and aviation authorities do not recognize the Bermuda Triangle as a special risk area. Ships and aircraft routinely travel through this region without incident, and there is no official warning or restriction for navigation within its boundaries.

Cultural Impact:

The Bermuda Triangle has become a prominent part of popular culture, featured in books, movies, and documentaries. While it continues to capture public fascination, the mystery surrounding the Bermuda

Triangle is often considered more myth than reality. The area within the Triangle is also susceptible to natural hazards common in any busy shipping and air traffic region, such as tropical storms, hurricanes, and unpredictable weather conditions.

The Bermuda Triangle has gained notoriety due to alleged mysterious disappearances. However, scientific analysis and investigation have largely debunked the notion of the Bermuda Triangle as a paranormal or hazardous area. The mystery surrounding this region is more a product of myth, speculation, and sensationalism than a scientifically validated phenomenon.

TOILET MUSEUM.........By Bindeshwar Pathak!

The Toilet Museum, also known as Sulabh International Museum of Toilets, was founded by Dr. Bindeshwar Pathak, an Indian social reformer and the founder of Sulabh International. The museum is located in New Delhi, India, and it is dedicated to the history and evolution of toilets and sanitation practices around the world.

Founder - Bindeshwar Pathak:

Dr. Bindeshwar Pathak is a sociologist and an advocate for sanitation and hygiene. He founded Sulabh International, an organization focused on promoting sanitation and providing affordable and sustainable sanitation solutions.

Dr. Pathak has been a prominent figure in the sanitation movement in India and has received several awards for his contributions, including the Stockholm Water Prize.

Sulabh International Museum of Toilets!

The museum was established in 1992. It is housed in Sulabh Bhawan, the headquarters of Sulabh International in New Delhi.

The primary objective of the museum is to educate visitors about the historical and cultural aspects of sanitation and toilets. It showcases the evolution of toilets, sanitation technology, and hygiene practices from ancient times to the modern era.

Exhibits:

The museum features a diverse collection of exhibits, including replicas of ancient and medieval toilets, information on the development of toilet technology, and displays on sanitation practices in different cultures.

It also includes exhibits on the role of toilets in public health, the importance of sanitation in preventing diseases, and the impact of inadequate sanitation on society.

Educational Purpose:

The Toilet Museum serves as an educational resource to raise awareness about the importance of sanitation and proper toilet facilities. It aims to break taboos associated with discussions about toilets and promote a healthier and cleaner environment.

Global Perspective:

The museum provides a global perspective on sanitation by showcasing toilets and sanitation practices from various countries and time periods. This helps visitors understand the diversity of approaches to sanitation across different cultures.

Recognition:

The Sulabh International Museum of Toilets has gained recognition for its unique and informative approach to addressing sanitation issues. It has been visited by people from around the world, including students, researchers, and tourists.

While the concept of a toilet museum may seem unusual at first, the Sulabh International Museum of Toilets plays a crucial role in advocating for improved sanitation and hygiene practices. It highlights the historical context of sanitation and emphasizes the need for accessible and sustainable sanitation solutions to improve public health globally.

VENETIAN MASKS..........Freedom and Transgression!

The Venetian masks have a rich history deeply rooted in the culture and traditions of Venice, Italy. These masks have been a significant part of Venetian life and celebrations for centuries, with a unique symbolism that encompasses freedom, transgression, and a sense of mystery!

Historical Context:

Venetian masks date back to the 13th century and were initially worn during the Carnival of Venice. The Carnival was a time when social norms were relaxed, and people from different social classes could interact freely and anonymously.

Freedom of Expression:

The use of masks during the Carnival allowed individuals to express themselves freely without fear of social judgment or consequences. The anonymity provided by the masks created a sense of liberation, enabling wearers to engage in behaviors or express emotions that might be restricted in their daily lives.

Social Equality:

Venetian masks played a role in promoting social equality during the Carnival. When everyone was disguised behind a mask, distinctions of class, gender, and social status were temporarily erased. This fostered a spirit of inclusivity and unity among the participants.

Transgression and Subversion:

The anonymity offered by Venetian masks allowed people to engage in activities that were considered transgressive or subversive. This could include political discussions, romantic encounters, or acts that challenged societal norms. The masks became a tool for temporary rebellion against the constraints of social expectations.

Symbolism and Mystery:

Venetian masks are often adorned with intricate designs and colors, adding an element of mystery. The concealment of the wearer's identity and emotions contributes to the allure and fascination surrounding the masks. This symbolism of mystery adds an extra layer to the freedom and transgression associated with their use.

Theater and Performance:

Venetian masks are closely tied to the theatrical tradition of Commedia dell'arte. The use of masks in performances allowed actors to play multiple roles and adopt various personas, challenging conventional storytelling and characterizations.

Modern Significance:

While the Carnival of Venice continues to be a major event for Venetian masks, these masks have transcended their historical context. They are now popular symbols in various cultural events, parties, and artistic expressions worldwide. In modern contexts, the masks may still represent freedom of expression and a departure from conventional norms.

The Venetian masks embody a sense of freedom and transgression deeply rooted in the historical and cultural context of Venice. They have served as symbols of liberation, social equality, and mystery, allowing wearers to temporarily break free from societal constraints and embrace a more liberated and expressive self.

PABLO ESCOBAR..........The Godfather!

Pablo Escobar, born on December 1, 1949, in Rionegro, Colombia, rose to infamy as one of the most notorious and powerful drug lords in history. From his early criminal activities to his rise as the leader of the Medellín Cartel and ultimately his death, Pablo Escobar's life is a tale of criminal enterprise, wealth, and violence…A true Godfather!

Early Life and Criminal Beginnings:

Escobar grew up in a lower-middle-class family. He was the third of seven children. His father worked as a farmer, and his mother was a schoolteacher. Early on, he showed an aptitude for illicit activities, engaging in small-scale criminal enterprises like stealing tombstones for resale.

In the late 1960s, he entered the drug trade by working for various criminal organizations, learning the intricacies of smuggling and distribution.

Formation of the Medellín Cartel:

In the late 1970s, Escobar, along with other drug traffickers, formed the Medellín Cartel. The cartel became a dominant force in the cocaine trade, controlling a significant portion of the global market.

Escobar was known for his ruthless tactics, which included violence against rivals, law enforcement, and civilians. The cartel's influence reached not only within Colombia but also internationally.

Wealth and Power:

At the height of his power, Escobar was one of the wealthiest individuals globally. His drug empire generated enormous profits, allowing him to live a lavish lifestyle and make significant contributions to Colombian society through charitable acts. His lifestyle was marked by opulence, including expensive cars, yachts, and a private fleet of planes. At his peak, Escobar was earning $5 Million per day and was considered the richest person in the world.

However, his wealth came at a great cost, as the drug trade fueled violence, corruption, and social instability.

Manhunt and Escapes:

Escobar's criminal activities attracted the attention of law enforcement and international authorities. The United States and Colombian governments, along with various law enforcement agencies, launched a massive manhunt to capture him.

Escobar managed to evade capture several times, often with the help of a network of loyal associates and sophisticated escape routes.

Support for Political Candidates:

Escobar, through his wealth and connections, supported various political candidates at local and national levels. He provided funding for their campaigns and expected loyalty in return. This support aimed to ensure politicians sympathetic to his cause were in power, offering protection against extradition and law enforcement efforts.

Attempts to Negotiate with the Government:

As pressure mounted on him, Escobar attempted to negotiate with the Colombian government. In the early 1990s, he proposed a deal known as "La Catedral Agreement," which involved surrendering to authorities and serving a short prison sentence in a facility of his choosing. However, this move was criticized for providing Escobar with significant privileges.

Political Party Formation:

Escobar went further in his political ambitions by attempting to create his own political party. In 1982, he founded the "Movimiento de Renovación

Liberal" (Movement of Liberal Renewal) party, known as MRL. The party aimed to represent the interests of drug traffickers and sought to gain political power.

Violent Tactics:

Escobar was not hesitant to use violent tactics to achieve his political objectives. He orchestrated bombings, assassinations, and other acts of terrorism to exert pressure on the government and manipulate political outcomes.

Impact on Colombian Politics:

Escobar's influence on Colombian politics had far-reaching consequences. It contributed to political corruption, erosion of institutions, and a weakening of the rule of law. The extent of his impact on the political landscape underscored the challenges faced by Colombian authorities in dealing with powerful drug cartels.

Ultimately, Pablo Escobar's political ambitions were intertwined with his criminal activities. His efforts to influence Colombian politics were marked by corruption, violence, and a willingness to exploit the political system to protect his interests. The legacy of his involvement in politics continues to shape discussions on the intersection of criminal enterprises and political power in Colombia.

Marriage:

He married Maria Victoria Henao in 1976 when he was 27, and she was 15. The couple had two children, a son named Juan Pablo and a daughter named Manuela.

Philanthropy and Populist Image:

Escobar strategically engaged in philanthropy to cultivate a positive public image. He built housing for the poor, constructed soccer fields, and invested in schools and hospitals in impoverished neighborhoods. These efforts earned him a reputation as a populist figure among some segments of the Colombian population.

Narcoterrorism and Violence:

Despite his public image as a benefactor, Escobar was responsible for numerous acts of violence. He orchestrated bombings, assassinations, and acts of terror to further his criminal enterprise and manipulate the political landscape. The Medellín Cartel, led by Escobar, was known for its ruthlessness in eliminating rivals and law enforcement officials.

Relationship with the Community:

While some people appreciated Escobar's charitable activities, others were critical, recognizing that his wealth came from the drug trade, which had devastating consequences for society.

Attempts to Shape Public Opinion:

Escobar was keen on shaping public opinion in his favor. He engaged with the media, providing interviews and presenting himself as a Robin Hood figure who sought to uplift the poor. His efforts were aimed at gaining sympathy and protection from the public.

Death and Legacy:

On December 2, 1993, after being on the run for several months, Pablo Escobar was located and killed by Colombian police in a rooftop shootout in Medellín.

His death marked the end of an era, but the impact of his criminal empire lingered. The Medellín Cartel's legacy continued to influence drug trafficking in the region and beyond.

Pablo Escobar's legacy remains controversial. While some remember him as a Robin Hood figure who provided for the poor and invested in local communities, others see him as a ruthless criminal responsible for countless deaths and societal chaos.

Pablo Escobar's life was marked by criminal activities, immense wealth, and a level of power that made him a central figure in the global drug trade. His story is a complex narrative that involves both admiration and condemnation, reflecting the profound and lasting impact of his actions on Colombia and the international community.

"Everyone has a price, the important thing is to find out what it is. There can only be one king. I can replace things, but I could never replace my wife and kids. All empires are created of blood and fire"

Pablo Escobar.

PEEPING TOM..........The Ultimate Punishment!

The legend of the Peeping Tom is a folklore tale that has been passed down through the centuries. The story is often associated with the city of Coventry in England and is commonly linked to Lady Godiva, a historical figure who, according to legend, rode naked through the streets of Coventry to protest her husband's oppressive taxation. The Peeping Tom character is said to have watched Lady Godiva's ride and was subsequently punished for his voyeuristic actions.......

Lady Godiva's Protest:

Lady Godiva, an 11th-century noblewoman, was married to Leofric, the Earl of Mercia. According to the legend, Leofric imposed heavy taxes on the people of Coventry. In response to the citizens' pleas for relief, Lady Godiva appealed to her husband to reduce the taxes.

The Condition:

Leofric, skeptical of Lady Godiva's chances of success, agreed to lower the taxes on the condition that she would ride naked through the streets of Coventry. Some versions of the story suggest that he expected her to refuse, but Lady Godiva took up the challenge.

The Ride:

Lady Godiva rode through the streets of Coventry on horseback, her long hair covering her body and preserving her modesty. The townspeople, respecting her sacrifice, reportedly stayed indoors, and only one person disobeyed.

Peeping Tom's Curiosity:

According to the legend, a man named Tom the tailor, later known as Peeping Tom, couldn't resist the temptation to witness Lady Godiva's naked ride. Despite the warnings and the agreed-upon privacy, he drilled a hole in his window shutters to observe the procession.

The Ultimate Punishment!

Lady Godiva, aware of the town's agreement to avert their gaze, was angered when she discovered that Tom had violated the pact. As a result, Peeping Tom was said to have been struck blind or dead either by divine intervention or the angered people of the town for his voyeuristic actions.

Lady Godiva's ride, along with the punishment of Peeping Tom, has become a famous legend. It is often cited as a symbol of a person's willingness to take a bold stand for justice, even in the face of personal sacrifice. The story has been retold in various forms over the centuries, including in literature, art, and popular culture.

Historical Basis:

While Lady Godiva was a real historical figure, the details of her legendary ride and the existence of Peeping Tom are considered to be more myth than historical fact. The tale has likely evolved over time, blending elements of folklore and moral lessons.

Cultural Impact:

The legend of Lady Godiva and Peeping Tom has endured and become a part of cultural lore. It has been referenced in literature, art, and even used as a term ("Peeping Tom") to describe someone who observes others without their knowledge or consent.

While the story of Lady Godiva and Peeping Tom has undoubtedly captured the imagination of many, it is important to recognize its mythical nature and its role as a cultural tale with moral undertones.

THE STAR FESTIVAL..........Orihime and Hikoboshi!

The Star Festival, known as Tanabata in Japan, is a traditional celebration that originates from Chinese folklore and has been embraced by Japanese culture. The festival typically takes place on the seventh day of the seventh month, according to the lunar calendar (which usually falls in early August in the Gregorian calendar). One of the central themes of Tanabata involves the love story of Orihime and Hikoboshi, two celestial beings separated by the Milky Way. Here is the story and significance of Tanabata:

The Story of Orihime and Hikoboshi:

Orihime is represented by the star Vega, and Hikoboshi is represented by the star Altair. According to the legend, Orihime was a skilled weaver, and Hikoboshi was a cowherd. They were deeply in love with each other.

Neglecting Their Duties:

Orihime and Hikoboshi were so enamored with each other that they neglected their responsibilities. Orihime's weaving suffered, and Hikoboshi's cows roamed freely. The heavenly king, Tentei, became displeased with their neglect and decided to separate them.

The Milky Way Separation:

Tentei separated Orihime and Hikoboshi by placing them on opposite sides of the Milky Way, a vast river of stars. They were allowed to meet only once a year on the seventh day of the seventh month if they worked diligently in their respective roles.

Tanabata Celebrations:

Tanabata celebrations involve people writing wishes on strips of colored paper called tanzaku. These wishes are then hung on bamboo branches. The bamboo branches, along with other decorations, are displayed in homes and public spaces during the festival.

Festive Decorations:

Festive decorations include paper lanterns, origami, and various ornaments. The streets of towns and cities are adorned with colorful displays, creating a vibrant and joyful atmosphere.

Wish-Making Traditions:

People use Tanabata as an opportunity to make wishes for the future. They write their hopes and dreams on tanzaku, hoping that their wishes will come true. These wishes are often for personal growth, success, good health, and happiness.

Regional Variations:

While Tanabata is widely celebrated across Japan, different regions may have unique customs and variations of the festival. The customs and specific date of the celebration may differ based on local traditions.

Modern Celebrations:

In modern times, Tanabata has evolved into a popular and colorful festival celebrated with parades, fireworks, and community events. The story of Orihime and Hikoboshi continues to be a central theme, emphasizing themes of love, perseverance, and the pursuit of dreams.

Tanabata is a beautiful and poignant celebration that blends ancient mythology with contemporary festivities. The legend of Orihime and Hikoboshi serves as a reminder of the enduring power of love and the importance of balance in one's life

YULE LADS.........The Miscreants!

The Yule Lads, or "Jólasveinarnir" in Icelandic, are mischievous figures from Icelandic folklore who are said to visit children in the thirteen nights leading up to Christmas. Each Yule Lad has his own distinct characteristics and behaviors, and they are considered a blend of playful mischief and somewhat ominous antics. The Yule Lads are often portrayed as the sons of Grýla, a mythical giantess, and her husband Leppalúði.........

Number and Names:

There are thirteen Yule Lads, each with a unique personality and mischievous trait. Their names, often translated into English, include Stekkjastaur (Sheep-Cote Clod), Giljagaur (Gully Gawk), Stúfur

(Stubby), Þvörusleikir (Spoon-Licker), Pottaskefill (Pot-Scraper), Askasleikir (Bowl-Licker), Hurdaskellir (Door-Slammer), Skyrgámur (Skyr-Gobbler), Bjúgnakrækir (Sausage-Swiper), Gluggagægir (Window-Peeper), Gáttaþefur (Doorway-Sniffer), Ketkrókur (Meat-Hook), and Kertasníkir (Candle-Stealer).

Arrival and Departure:

The Yule Lads begin to arrive one by one on each of the thirteen nights leading up to Christmas, starting from December 12th. They leave one by one in the same order, with the last one departing on January 6th.

Mischievous Behavior:

The Yule Lads are known for engaging in playful but mischievous activities. For example, Stekkjastaur is known for harassing sheep, while Þvörusleikir licks wooden spoons, and Skyrgámur steals skyr (Icelandic yogurt). Each Yule Lad has a unique form of mischief associated with their name.

Historical Origins:

The Yule Lads have roots in Icelandic folklore dating back to at least the 17th century. They were traditionally portrayed as figures who would either reward or punish children based on their behavior during the Christmas season.

Evolution of Tradition:

In modern times, the Yule Lads have become more lighthearted and are often portrayed in a playful and entertaining manner. They have become a beloved part of Icelandic Christmas traditions, with their images appearing in decorations, stories, and festivities.

Association with Grýla:

The Yule Lads are said to be the sons of Grýla, a giantess who is also part of Icelandic folklore. Grýla is often depicted as a frightening character who kidnaps and eats misbehaving children. The Yule Lads, while mischievous, are not as sinister as their mother.

Modern Celebrations:

In contemporary Icelandic culture, the Yule Lads have become a popular and endearing part of the Christmas season. Families often celebrate the arrival of each Yule Lad, and children leave their shoes out in the hope of receiving small gifts or treats.

The Yule Lads represent a unique and playful aspect of Icelandic Christmas traditions. While mischievous in nature, they have evolved from figures associated with potential punishment to beloved characters that bring joy and entertainment during the holiday season.

NOTE!

This Book is not devoid of errors. Proper research for accurate information is encouraged. The content presented in this book is intended for educational purposes, aiming to provide insights and stimulate thoughtful discussions on diverse subjects. All rights reserved. No part of this book may be reproduced, distributed, or transmitted in any form or by any means, including photocopying, recording, or other electronic or mechanical methods, without the prior written permission of the publisher, except in the case of brief quotations embodied in critical reviews and certain other noncommercial uses permitted by copyright law.

www.ingramcontent.com/pod-product-compliance
Lightning Source LLC
Chambersburg PA
CBHW050441290526
45786CB00006B/2109